Japanese-English bilingual book

日本語-英語バイリンガル・ブック

Design Science for
Product Creation × Product Usage

10 Guidelines for Creating New Values
Toward Business Strategy

モノづくり×モノづかいのデザインサイエンス

経営戦略に新価値をもたらす10の知恵

松岡由幸 著
Yoshiyuki Matsuoka

近代科学社
Kindai-Kagaku-Sha

はじめに

　急激に進化する AI, IoT, ビッグデータ. それらを背景にして進む, インダストリー 4.0 に向けたさまざまな動き —— このように, 今, モノづくりは大きな変革のなかにある. さらに, これらのモノづくりには, モノの使用段階におけるサービスビジネスとのシナジーが期待されている. いわゆる「モノづくり×モノづかい」の新たなビジネスモデルへの転換である. このように, 今, モノづくりとモノづかいは激動の時代を迎えている. そして, それらの経営戦略には, 新たな価値の創造とそれをもたらす知恵が不可欠であることは言うに及ばない.

　そこで, 本書では, 近年, 急速に発展してきた「デザインサイエンス」に注目する. 本書でいう「デザイン」とは, 経営戦略, 商品企画, 技術開発を含むモノづくりとモノづかいを指す. 近年, 「デザイン思考」が経営にも取り入れられるようになっている. 本書では, その思考を含み, それの生みの親である「デザインサイエンス」がもたらす未来創造の"10"の知恵を紹介する. これらの知恵に, 新価値創造と超競争社会に勝ち続ける企業の頑強性構築のヒントを, 読者の皆様が見いだしていただければ幸いである.

PREFACE

AI (Artificial Intelligence), IoT (Internet of Things), and big data are on the process of remarkable development. At the same time, various movements toward Industry 4.0 are in progress— In fact, product creation is currently in the period of significant revolution. Along with this period, the synergy between the product creation and the business of services applied to the usage process is being expected. In other words, the product creation is shifting over to a new business model of "product creation × product usage". Therefore, the both product creation and product usage are facing the period of significant change. It is needless to say that the creation of new values and its how-to knowledge are indispensable for the business strategy.

This book focuses on the developing field "design science". Here, the word "design" is used as the act of not only the product creation (e.g., building a business strategy, planning a product concept, developing technology) but also the product usage. As the matter of fact, in recent years, "design thinking" —one of the design concepts of "design science"— is being integrated to various business. In this book, 10 guidelines offered by "design science", and how they will contribute to your business future are introduced. The knowledge will hopefully provide you with various hints to bring about meaningful toughness within your organization, and to be successful in creating new values toward your business strategy.

目　次

はじめに ……………… ii

第1の知恵	「感動」を生みだす ── 心を動かすモノづくりのために	2
第2の知恵	「つかう」をつくる ──「モノづくり」×「モノづかい」の産業化に向けて	16
第3の知恵	「価値成長」を仕掛ける ── 使い捨て社会からの脱却に向けた， 　　新たなモノづくり	28
第4の知恵	「タイムアクシスデザイン」で拓く ── 独自の産業化と経済再生に向けて	44
第5の知恵	「AGE思考」で思いつく ── 本当に，AIは人間の創造力を超えるか？	58
第6の知恵	「創発」に学ぶ ── 人はなぜ創造できるのか？	76
第7の知恵	「状態」を考える ── 製品のデザインから関係性のデザインへ	90
第8の知恵	「ロバスト性」を獲得する ── 強靭で，持続可能なモノづくりと社会のために	104
第9の知恵	「多空間デザインモデル」で開発する ── Mメソッド：新たな価値創造に向けて	116
第10の知恵	「マルチコンカレント」でリードする ── 第4次産業革命がもたらす 　　「早い者勝ち時代」に何をなすべきか？	130

付録 ……………… 146
　1. デザインサイエンス
　2. 多空間デザインモデル（Mモデル）
　3. AGE思考モデル

参考文献 ……………… 160
索引 ……………… 162
著者紹介 ……………… 166

CONTENTS

PREFACE ······· iii

1st Guideline	**Arousing "Deep Emotion"** — In Order to Create Affecting Products	3
2nd Guideline	**Creation of "Usage"** — Toward Industrialization of "Product Creation" × "Product Usage"	17
3rd Guideline	**Plant of "Value Growth"** — New Product Creation for Breakaway From Throw-Away Society	29
4th Guideline	**Adapting "Timeaxis Design" to Development** — Toward Original Industrialization and Economic Revitalization	45
5th Guideline	**Idea Generation Using "AGE Thinking"** — Will AI Surpass Human Creativity?	59
6th Guideline	**Learning from "Emergence"** —Why Can Humans Create?	77
7th Guideline	**Considering "State"** —From Product Design to Relationship Design	91
8th Guideline	**Ensuring "Robustness"** — For Tough and Sustainable Manufacturing and Society	105
9th Guideline	**Applying "Multispace Design Model" to Innovation** — M Method: Toward New Value Creation	117
10th Guideline	**Leading by "Multi-Concurrent"** — What Should We Do in this Period of "First-Come, First-Serve" Brought by the 4th Industrial Revolution?	131

APPENDIX ······· 147
  1. Design Science
  2. Multispace Design Model: M Model
  3. AGE Thinking Model

REFERENCES ······· 161
INDEX ······· 163
AUTHOR NOTE ······· 167

第 **1** の知恵

「感動」を生み出す

心を動かすモノづくりのために

1st Guideline

Arousing "Deep Emotion"

In Order to Create Affecting Products

感動の不思議

あなたは，最近，どのようなことに**感動**されただろうか？ 映画？ コンサート？ スポーツ観戦？ それとも，友人との再会？ 家族との出来事？ このように，人はさまざまな場面で感動する．では，感動とは何なのか？ 広辞苑によると，感動とは「深く物に感じて心を動かすこと」とある．しかしながら，感動については諸説あるものの，実はあまり解明されておらず，いまだ不思議に満ちているのである．

そのようななか，著者らは，デザイン科学の立場から感動について研究を進めてきた．感動に関する知見や指摘事項を，哲学，美学，心理学，生理学，生物学などさまざまな領域から収集し，言語，画像，動作などさまざまな記号といわれるものの本質・在り方・機能を探究する記号論の視点から考察を行ってきた．

そこでここでは，感動にはどのようなタイプがあるのか，どのような条件がそろうと感動するのか，さらに，そもそも何のために人は感動するのかについて述べていきたい．さらに，今後の，人の心を動かすデザイン・モノづくりの在り方についても考察を加える．

Mystery of Deep Emotion

Have you felt **deep emotion** by anything recently? Is it a movie? Concert? Sports game? Or perhaps a reunion with your old friend? Maybe a special happening with your family? As you can see, people feel deep emotion and their heart can be affected in various scenes. But then, what exactly is deep emotion? In general, deep emotion is said as "feeling in which one deeply feels toward a thing and gets their heart affected". However, even though there are many opinions and views, the real mechanism of deep emotion is not yet solved and still left in mystery.

Under such present condition, we have been proceeding in the study about deep emotion from the design science point of view. We have collected knowledge and indication matter about the deep emotion from various fields such as philosophy, aesthetics, psychology, physiology, and biology. Then, we considered the semiotics perspective by understanding the essence and functions of symbols like language, images, and movement.

Therefore, in this article, I would like to introduce you the types of deep emotion, the condition necessary to achieve deep emotion, and the reason why people feel deep emotion. Furthermore, the consideration about the ways of future design and product creation which give deep emotion and affects one's heart, will be discussed.

感動の3タイプ

見方・切り口にもよるが，感動の種類にはどうやら，以下に示す三つのタイプが存在するようである（図1）．

一つ目のタイプは，美しい夕陽や風景，絵画，音楽など対象そのものに対する「**オブジェクト型**」の感動．これは，モノづくりにおいて，製品そのものの機能性や造形性などに対して生まれる感動である．この感動は，モノづくりが目指す基本であるともいえるのではないだろうか．

二つ目のタイプは，対象の背景にある能力や努力などを強く感じることで生まれる「**バックグラウンド型**」の感動．この感動は，自分では到底不可能な能力や努力などを感じる時に多く発生する．この感動においては，敬意や時には畏怖さえも感じることがあるようである．これは，モノづくりでいえば，製品そのものの評価を超え，その製品のメーカに対する評価につながる．そのため，ユーザーの安定的な信頼感や価値観を生みだすことから，モノづくりにとっては重要な感動といえるだろう．

三つ目のタイプは，友人や家族など人との関係においてよく現れるものであり，その対象（出来事）に出会うまでの文脈（経緯）が大きく関与する「**コンテクスト型**」の感動である．これは，さまざまな映画，コンサート，小説など時間軸を有する場合によく現れる．日本の花火も「コンテクスト型」といえる．花火が空高くドーンと開くまでの，地上からのヒュルヒュルヒュルー──この文脈が期待感を膨らませて感動を生んでいるのは，皆さんも実感しておられるのではないだろうか．

3 Types of Deep Emotion

From our study, there seems to be 3 types of deep emotion (figure 1). The first type is called the "**object type**", an impression felt toward the objects like beautiful scenery, artwork, and music. In product creation, this type of deep emotion arises from the functionalities and aesthetics characters of the products. It may be said that, this type of deep emotion is the basics the product creation is aiming at.

The second type of deep emotion is called the "**background type**", an impression felt toward abilities or efforts in one's (thing's) background. This type of deep emotion arises when people feel unachievable and impossible abilities or efforts. Moreover, it even seems to achieve awe and respect at times. In product creation, this type leads to the evaluation of not only the products, but also the makers of the products. Therefore, it may be said that this deep emotion is important in product creation, to achieving people's stable trust and sense of values.

The third type of deep emotion is called the "**context type**", an impression influenced from the contexts (processes) occurring to the objects (events). This type of deep emotion is seen in relationships of families and friends, and also in things including timeaxis such as movies, concerts, and novels. The fireworks of Japan are also included in the "context type". This is because their contexts as "the time in which the fireworks are lit until they are shot up to the sky to bloom" gives people high expectation and triggers their feeling of deep emotion.

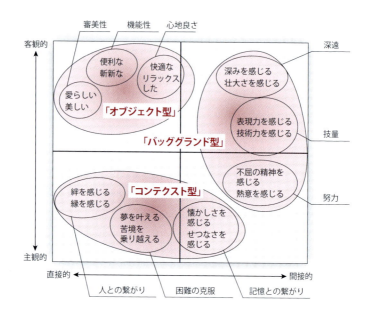

図1:感動の3タイプ

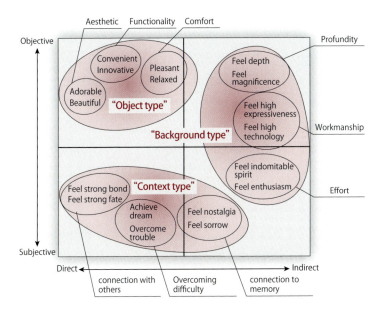

Figure 1 : 3 types of deep emotion

1st Guideline : Arousing "Deep Emotion"

深い感動のために

「コンテクスト型」の例として，私は，以前に観た上海雑技団のアクロバットを思い出す．一人の男の子が別の男の子の肩の上に飛び乗る．次に，その男の子の肩に，さらに別の男の子が飛び乗る．そして，次々と……．6人目の子が飛び乗ろうとし，何度か失敗するも何とか成功．そして，7番目の子がトライする．何度も何度も失敗を繰り返す．さすがにもう駄目かなと思った矢先，まさかの成功．会場は拍手喝采となった．このような感動的なアクロバットを，皆さんも観られたことがあるのではないだろうか．

しかし，この感動の事例をよく見ると，これは失敗が重なる単なる「コンテクスト型」だけではないことがわかる．7人の男の子が高く重なる「オブジェクト型」としてのすばらしさ，さらに「バックグラウンド型」としてのその背景にある莫大な努力．このアクロバットは，それらをすべて兼ね備え，同時に感じることで生まれた感動であるといえる．実際の感動においては，このような複合型により，深い感動につながる場合が多いのも事実であろう．

For Deeper Emotion

As an example of the "context type", I remember the Shanghai acrobatics groups I watched before. One boy jumps on the shoulder of a different boy. Then, another boy jumps onto the shoulder of the boy. And then another boy again and again... At the sixth boy, they failed but kept on trying and finally succeeded. Then, the seventh boy repeatedly fails and the hopes of the audiences were low. However, at the last moment, the seventh boy succeeded and impressed all the audiences. Perhaps everybody might have seen such impressive acrobats.

This example is not just a simple "context type" for undergoing several failures. It also has the splendor of the seven boys, as the "object type", and the boys' hard efforts to the act, as the "background type". Consequently, it can be said that the Shanghai acrobats had all types' factors of deep emotion, and the audiences were impressed by the aforementioned appearances of the boys. In fact, the deep emotion arose in reality contains 2 or 3 types' factors, which often lead to deeper emotion.

感動のメカニズム

では，これらの感動に共通する要因は何であろうか？ 著者らの研究によると，その要因は「**共感**」と「**驚き**」であることがわかってきた．「共感」と「驚き」の両方が備わった際にはじめて，感動が生まれるのである．このことは，機能性や使いやすさに共感を生む新製品から驚きを伴う現代アートまで，さまざまなモノを対象に心理実験を行い検証されている．この**感動のメカニズム**を表す論理式を，図2に示す．

ここで，「共感」と「驚き」の関係に注目する．一般に，慣れ親しんだことには共感しやすい．しかし，慣れ親しんだことには驚かない．つまり，「共感」と「驚き」は二律背反の関係にあり，図2に示す少ない共通部分で人は感動することになる．そのため，感動的なモノづくりを行う際には，この少ない共通部分を見つけ出し，それを具現化するという難しさと，そして面白さがあるといえる．

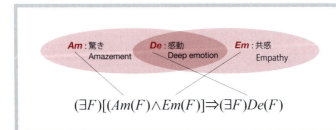

Mechanism of Deep Emotion

Then, what is the common factor of these deep emotion? From our studies, we came to an understanding that "**empathy**" and "**amazement**" are the factors that arouse deep emotion. In other words, when "empathy" and "amazement" are achieved, the deep emotion is aroused for the first time. This fact was proved through a psychological experiment on various things such as new functional products and modern art which one can be empathized or amazed with. The logical formula of the **mechanism of deep emotion** is shown in figure 2.

Here, we will pay attention to the relation of "empathy" and "amazement". In general, people will empathize with things they are familiar with. However, they will not get amazed to things they are familiar with. In other words, "empathy" is in a relation of antinomy with "amazement", and people will feel deep emotion only in the small common parts of the two factors (figure 2). For that reason, in order to create products which would arouse deep emotion (affect one's heart), we must find out the rare common parts between "empathy" and "amazement", which is not only a difficult but also an interesting task to achieve.

図2：感動の論理式

Figure 2 : Logical formula of deep emotion

心を動かすモノづくり

　モノづくりにおいても，「共感」と「驚き」を伴う感動がほしいところである．そして，その感動の生起には，それまでにユーザーがどのような使用経験をしたかが大きく左右する．

　近年，デザイン科学の領域において，ユーザーの経験をデザイン対象とする**ユーザーエクスペリエンスデザイン**（以下，**UXデザイン**）や**タイムアクシスデザイン**などが，盛んに議論されている．UX デザインとは，ユーザーがある製品やシステムを使ったときに得られる経験や満足などをデザインすることである．また，タイムアクシスデザインは，文字通り，人が使用する経験などの時間軸をデザインすることである．これらのデザインは，いずれもユーザーの経験とそこから生まれる価値を操作しうる．そのため，先述した感動のタイプやメカニズムの知見を応用しつつこれらのデザインを行うことで，感動の生起をうまく引き出せる可能性が十分にあると考える．

　今後，モノづくりにおいては，ますます人の心を動かす感動が望まれている．そして，そのためのヒントが UX デザインやタイムアクシスデザインに潜んでいるものと考える．

Product Creation That Arouses Deep Emotion

It is desirable to arouse deep emotion with both "empathy" and "amazement" in product creation. Also, occurrence of deep emotion often depends on the people's experience of how they have used the product before.

In recent years, design intended toward the user's experience, such as **user experience design** (**UX design**) and **timeaxis design** (**TaD**) is being discussed in the field of design science. UX design is to design the experience or the satisfaction the user will experience when using the product or the system. In addition, TaD is to design the timeaxis of the user experience. These designs can operate the user experience and the value born from the experience. Therefore, it is believed that by applying the 3 types of deep emotion and their mechanism to these design, will enable the possibility of arousing deep emotion.

In the future product creation, arousing deep emotion will be increasingly expected. The hint to realizing such design might lie in UX design and TaD.

第 2 の知恵

「つかう」をつくる

「モノづくり」×「モノづかい」
の産業化に向けて

2nd Guideline

Creation of "Usage"

Toward Industrialization of
"Product Creation" × "Product Usage"

「つくる」と「つかう」の分化

人類が道具をつかいはじめた石器時代のころ,人はモノ(道具)を自らがつくり,自分でつかっていた.つまり,「つくる」人と「つかう」人が同一であった.しかし,やがてモノづくりに長けた名人が現れると,状況は大きく変化する.その名人が「つくる」を一手に担い,「つくる」人と「つかう」人の分化が始まった.そして,この分化は,18世紀後半の産業革命以降さらに顕著となり,今日に至っている.

現在の「モノづくり」では,まず,「つくる」人が「つかう」人のことを十二分に考えて「モノ」をデザインし,生産する.次に,「つかう」人がその生産物をつかい,「機能」を実感する.その際,「機能」が不十分であれば,「つくる」人はその改良を図り,再び生産する.そして,「つかう」人はまたつかってみる.この循環システムが今日の「**モノづくり**」の基本となっている.

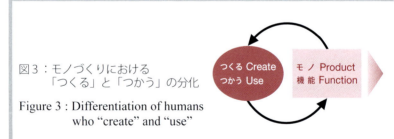

図3:モノづくりにおける「つくる」と「つかう」の分化

Figure 3 : Differentiation of humans who "create" and "use"

Culture of "Creation" and "Usage"

In the days of the Stone Age, when the humans have begun to use tools, the humans created products (tools) by themselves and used it by themselves. In other words, the person to "create" and "use" were the same. However, the situation greatly changes, when an expert of product creation appears. The expert becomes the one to "create", and this has started the differentiation of the humans who "create" and "use". The differentiation of people becomes more remarkable in the late 18th century and continues up to present day.

In current product creation, first the person who "create" thoroughly thinks about the people who "use", then design a product, and produce the product. Next, the people, who "use" the product will use it, and experience a function of the product. If the function is insufficient, the people who "create" will make improvements and reproduce the product. Then, the people who "use" the product will again, use the improved product. This circulation system is the bases of today's "**product creation**".

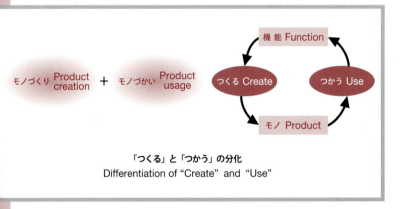

「つくる」と「つかう」の分化
Differentiation of "Create" and "Use"

機能は「つかう」場に依存する

　ここで，モノの機能について改めて考えてみる．

　たとえば，おいしいお酒をいただくために，人は，好きな銘柄のお酒を選ぶだけではない．肴やぐい呑み，酒屋の雰囲気，呑み仲間，そして呑み方など，さまざまな「**場**」も大事であり，それらはおいしさに大きく影響する．それと同様に，モノの機能も，モノの特性だけでは決まらない．モノの特性とそれが使用される場の特性との組み合わせで決定する．

　この視点に立てば，機能を高める手段としては，モノを「**つくる**」**デザイン**だけでなく，それを「つかう」場のデザイン（以下,「**つかう**」**デザイン**）も有効ということになる．しかしながら，これまでは，どちらかといえば，前者の「つくる」デザインが主流であり，後者の「つかう」に関しては，その調査や分析の結果を，あくまで「つくる」デザインの条件として利用するにとどまることが多かった．しかし，それではもったいないと思うのである．

「つかう」デザインに注目

　そこで，「つかう」デザインに注目する．このデザインは，従来型の「つかう」ことを考慮してデザインすることではなく，つかい方やつかわれる環境といった「場」それ自体をデザイン

Function Depends on Circumstances to "Use"

Here, we will discuss about the function of the product. For example, in order to drink a delicious liquor, one must not only consider about the liquor brand. One must also consider about various "**circumstances**" such as, side dishes, liquor glasses, atmosphere, drinking friends, and ways of drinking which affects the taste of the liquor. The function of a product is similar to this example, and cannot be decided only with the characteristics of the product. It is the combination of the product's characteristic and the usage circumstances of the product that decides its function.

From the perspective of considering the circumstances of the function, it could be said that other than the **design for "creation"**, the design of the product "usage" (**design for "usage"**) is effective. However, the design for "creation" have been the most popular up until now, and the design for "usage" was only used as the condition of the design for "creation" through the research and analysis of the usage. Furthermore, I think that not only the design for "usage" but also the design of "usage" must be focused.

Paying Attention to Design of "Usage"

Therefore, the design of "usage" is put into consideration. The design of "usage" discussed from here is not the former design for "usage", but a design which designs the

することを指す．その意味で，「**モノづくり**」のデザインではなく，「**モノづかい**」のデザインといえるのかもしれない．

　たとえば，環境問題に関していえば，すでに「つくる」だけの対応では限界がきているといえるのではないだろうか．自動車の排ガス対策の多くは，従来，エンジン改良などの自動車を「つくる」技術に依存する傾向にあった．しかし，どのような道路環境でいかにペダル操作を行うかといった「つかう」技術を適切に用いることが，大きな効果につながることは周知である．このような場に関する技術を応用する「モノづかい」のデザインに関しては，いまだ多くの可能性が残されているのではないだろうか．

「つくる」と「つかう」の相乗効果

　「つかう」デザインは，さまざまである．たとえば，つかいやすさの向上のために，場を構成するモノ（道具）をデザインする場合も多い．これは，「モノづかい」のための「モノづくり」であるともいえ，操作系や表示系のデザインとしてはすでにいくつか見受けられ，いずれも大変興味深いものである．

　しかしながら，ここでいう「つかう」デザインは，「モノ」のデザインだけでなく，そのつかい方など「コト」に関するサービスのデザインである．たとえば，ピアノ製造業が行うピアノ教室や調律はその身近な一例であろう．ユーザーはピアノを購入するのみならず，ピアノを練習し，調律を施す．そして，その練習や調律は，音の質を高めてくれる．さらに，ユーザーに

"circumstance" such as, usage or usage environment of the product. From this definition, it could be said that this design is for "**product usage**" and not for "**product creation**".

As an example, the design for "creation" has come to its limits to correspond to the environmental problems. Most smug emission countermeasures for vehicles, have been dependent on design for "creation" (e.g. engine improvements). However, it is well-known that the appropriate application of the "usage" technique (e.g. understanding how the pedal operation is performed under road conditions) leads to a huge effect. The "product usage" design, which applies the technology considering the circumstances, may still have numerous possibilities within it.

Synergy of "Creation" and "Usage"

There are many varieties in design of "usage". For example, there are lots of cases to design a product (tool) that constitutes the circumstances, to improve its usability. This could be said as "product creation" for "product usage", and is already found in operation and indication system, which are both very interesting.

Nevertheless, the design of "usage" described here, does not only indicate the design of a "product", but also the design of a service, including the usage of a product. The piano lesson and the tuning, the piano industry provides, can be given as the example. The user not only purchases the piano, but also practices and tunes it, which lead to raise the quality of the

はそのピアノに対する愛着が生まれ，そこには，まさに「つくる」と「つかう」の相乗効果が仕組まれている．

タイムアクシスデザイン

　タイムアクシスデザインも，「つくる」と「つかう」の相乗効果を狙うデザインの一例といえる．このデザインは，文字通り，時間軸をデザインすることである．具体的には，「つくる」デザインにおいて，生命システムの特徴を応用するインスパーヤード技術（ある種の制御技術を含む）をモノに組み込む．そうすることで，環境に適合した時間軸変化を可能にする．また，「つかう」デザインにおいては，サービス技術を用いる．それにより使用段階での使用条件や社会・価値観などの時間軸変化への対応を可能にする．タイムアクシスデザインは，これらの特徴により，つかえばつかうほど価値を成長させ長期使用を可能にすることで，使い捨て社会からの脱却やモノを大切にする精神文化の醸成も視野に入れている．このデザインは，「モノづくり」と「モノづかい」を合わせたビジネスを可能にし，**2次産業（モノづくり）×3次産業（モノづかい）を一体化させたもう一つの6次産業化**としての経済効果も期待されている．

sound. Furthermore, the attachment the user feels toward the piano grows through the practicing and tuning. In fact, there is synergy of "creation" and "usage".

Timeaxis Design

Timeaxis design is also one of the examples of design aiming for the synergy of "creation" and "usage". This design literally means to design the timeaxis. To be specific, the design of "usage" incorporates the bio-inspired technology that employs the life system's trait to a product. The bio-inspired technology enables timeaxis change of the product adapted to the environment. In addition, the service technology is applied to the design of "usage". The service technology enables the correspondence of timeaxis change, such as the changes of usage conditions and the changes of society and sense of values. The timeaxis design uses such traits to realize a product with long-term usage and also grow its value through continuous usage. Such design can break away from a throw-away society and construct mental culture, where every product is treated with respect. In addition, the timeaxis design enables a business that causes the synergy of "product creation" and "product usage", and unify the **secondary sector of industry (product creation)** with the **tertiary sector of industry (product usage)**, as the **6th industry**. This new industry is expected to bring high economic effect.

「つくる」と「つかう」の再考

　本章では,「つくる」と「つかう」の関係から,「つかう」デザインに注目した.さらに,「つくる」と「つかう」の相乗効果を生む「モノづくり」と「モノづかい」を一体化させた産業化についても言及した.

　近年,「つくる」と「つかう」の関係に,多様性が生まれつつある.タイムアクシスデザインは,その一例として先述した.また,**3Dプリンター**もその一つである.この機械は,ユーザーが自らつくり,自らつかうという「つくる」と「つかう」の新たな関係を可能にする.これは,産業革命以降の大量生産型のモノづくりに一石を投じるかたちとなっており,さまざまな議論がなされている.

　今,我々は,「つくる」と「つかう」について,再考すべき時を迎えているのかもしれない.また,その再考が,これからの新たなモノづくり産業へのヒントを生み出す契機になることを期待する.

Reconsideration of "Creation" and "Usage"

This chapter focused on the design of "usage" from the relation of "creation" and "usage". Furthermore, the industrialization of causing the synergy of "product creation" and "product usage", was discussed.

In recent years, the relations of "creation" and "usage" are diversifying. The timeaxis design was given as one of those examples. Also, **3D printer** is another example, because it enables to create a new relation of "creation" and "usage", through the process of a user creating and using a product by themselves. Various arguments are done along with the appearance of the 3D printer, since it has caused a stir in the mass production manufacturing, since the Industrial Revolution.

Perhaps, it may be time, for us to reconsider "creation" and "usage". In addition, this reconsideration is hoped to become the opportunity in gaining a hint to realize a new manufacturing industry toward our future.

第3の知恵

「価値成長」を仕掛ける

使い捨て社会からの脱却に向けた,新たなモノづくり

3rd Guideline

Plant of "Value Growth"

New Product Creation for Breakaway
From Throw-Away Society

新コンセプト「価値成長デザイン」

　自動車の開発に従事していた若いころ，私には一つの残念な思いがあった．どうして，自分たちのつくる自動車は，お客様が購入されたときに一番価値が高いのだろうか？　その価値は使用するにつれ，やがて価値が下がってくる．何とか，使えば使うほど機能性が高まり，愛着が増し，価値を高めていくことはできないものだろうか？　私は，そのようなモノづくりへの思いを馳せながら，長年，自動車開発に従事してきた．しかしながら，当時その具体的な策を見つけることができず，結局，価値を上げるデザインは実現しなかった．

　ところが，近年，デザインサイエンスの進展に伴い，ようやくそのヒントが見つかった．使うにつれ価値が成長するデザインの可能性が見えてきたのである．私は，そのような，使用段階において価値が高まり，長く使用可能な製品をデザインすることを，「**価値成長デザイン**」と呼んでいる．

　この価値成長デザインの特長は，製品の使用段階の時間軸変化に注目する点にある．その直接的な狙いは，製品とそれが使用される場の関係から決定される機能や価値を，時間軸で持続的に適正化しつづけることである．これにより，ニーズや使用環境に加え，ユーザーの価値観などの多様性に対して個別最適化を図ることができる．さらには，社会や生活の時間軸変動にも対応する．それにより，長期使用による環境負荷の低減や使い捨て社会からの脱却など，さまざまな効果を生むことが期待されている．

New Concept, "Value Growth Design"

I had this one disappointing thought back when I was young and still engaged in automobile development. I wondered, "why is it that the vehicles we make has the highest value to the customers the moment when it is purchased? After purchase, the value of the vehicle keeps on decreasing through usage. Is there not a way to increase the value of the vehicle through usage, so that the functionality and the attachment toward the vehicle rises?" I have engaged in automobile development for many years, with these thoughts in my mind. However, back in those days, I was not able to find a resolution to realize a design in which its value increases.

Nevertheless, the recent progress of design science has brought the clue to realize value growth design. I have found the possibility of a design, in which the value grows through usage. I have called such design, "**value growth design**", which allows design to increase the product's value through usage and to be used for a long term.

The characteristic of this value growth design is the point in which it is focusing on the timeaxis change of the use stage of the product. The direct aim of the value growth design is to continuously deciding the adequate value and function of the product by considering the relationship with the product's usage circumstances through timeaxis. The value growth design can allow individual optimization by corresponding to needs and use environments of the product, and sense of values and variety of the user. In addition, the value growth design can also correspond with the timeaxis change of the society and life. It is thereby, expected that the value growth design can provide various effect including, reduction of environmental load and breakaway from the throw-away society by realizing a long usage design.

価値成長の4タイプ

これまでの研究により，価値成長デザインには四つのタイプが存在することが明らかになっている．

・自然変化型：

使用に伴い，化学変化するタイプ．「**自然変化型**」の事例としては，漆器，南部鉄器，耐候性鋼，銅板屋根などが挙げられる．日本の美を代表する漆器は，その酸化反応により，時を経て徐々に硬化し丈夫になることがよく知られている．建築外板に用いられるコールテン鋼と呼ばれる耐候性鋼も注目されている．徐々に酸化により現れる錆は耐食性を向上させるとともに，美しい色合いを醸し出す．近年欧州で注目されている南部鉄器の急須で飲むお茶を，その錆がおいしくさせている．新たなところとしては，自己修復コーティング材による傷の治癒も，今後大きく期待される．価値成長の観点から考えると，この自然変化型材料はまだまだ，さまざまな可能性があるのではないだろうか．

自然変化型

外部環境との間で化学変化が起きることで，耐久性，性能が向上する

● 南部鉄器，漆器

図4.1：価値成長の4タイプ (1)

4 Types of Value Growth

Previous studies have clarified the existence of 4 types of value growth design.

・Natural Changing Type:

"**Natural changing type**" is a type of value growth design, in which durability and efficiency improves by chemical reactions with the product's surrounding environment. The nambu ironware, lacquerware, corduroy steel, copper-sheeted-roof are examples of a natural changing type. It is well known that the lacquerware that represents Japan's beauty, gradually stiffens by the oxidation reaction over time, making it stronger. In addition, the corduroy steel used for building shell is also attracting attention. The oxidation reaction gradually rusts the iron, which improves corrosion resistance and brings on a beautiful color. The recently popular nambu ironware is also making the tea more delicious by its rust. As a new factor, the self-repair coating materials that can heal the wound by itself, is highly expected in the future. From the value growth point of view, this natural changing type may have many possibilities more than we can imagine.

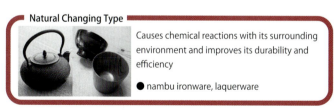

Figure 4.1 : 4 types of value growth artifacts (1)

- 馴染み型：

使用に伴い，形状や物性がユーザーに馴染むタイプ．「**馴染み型**」に関しては，筆，ゆがけ（弓道に用いる手袋），紬，万年筆（ペン先）などの事例が見受けられる．高級筆である豊橋筆は，使い込むほどに使う者の手に馴染み，一生ものとも言われている．筆職人によれば，この筆は，書道家の手の力，動かし方をも考慮してつくり込まれることで，はじめて実現されるという．そこには，使い手と作り手が対話してモノをつくるという，モノづくりの一つの理想が存在している．「馴染み型」の事例には，皮革，布，金属など材料の弾塑性や摩耗特性などの力学特性に関与するものが多い．そのため，今後の製品開発の視点としては，それらの材料特性に注目することで，さらなる新たな馴染み型価値成長デザインの可能性があるものと考える．

馴染み型

使用を重ね，手入れを行うことで，ユーザに馴染み個別に最適化する

● 革靴，万年筆

図4.2：価値成長の4タイプ（2）

• Accustoming Type:

"**Accustoming type**" is a type of value growth design, in which the product becomes accustomed personally to its user through repeated usage and care. The writing brush, archer's glove, pongee, fountainpen (pen point) are examples of the accustoming type. The highly expensive brush becomes accustomed to the user through repeated care, and is said to last for a life. According to calligraphy craftsman, the Toyohashi brush is truly realized, only if the user's hand strength and movement are considered when making the brush. This is one of the ideal product creation, in which the user and the designer interacts to create a product. The "accustoming type" has several examples such as leathers, cloths, and metals which holds a mechanical characteristic such as elastic-plasticity and attrition characteristics. Therefore, from the future product development point of view, considering the given material characteristics lead to a new accustoming type value growth design.

Figure 4.2 : 4 types of value growth artifacts (2)

・カスタマイズ型：

　使用に伴い，カスタマイズするタイプ．「**カスタマイズ型**」については，加工性の良好な木材や金属製の家具などは周知である．また，スマートフォン，カーナビなどのIT機器も，すでに多く市場に存在する．新しいところでは，調光ガラスを用いることで，ユーザー個別の嗜好や使用環境に応じた調光をカスタマイズする事例も出てきている．また，形状記憶の介護用スプーンなど，介護・医療現場において個別に最適な形状にカスタマイズする事例も現れており，この領域に関しては，今後の新たな展開が期待できる．

ユーザが独自のカスタマイズを施すことで，使用環境に適合する

●カーナビ，スマートフォン

図4.3：価値成長の4タイプ (3)

- Customizing Type:

"**Customizing type**" is a type of value growth design, in which the product adapts to its operating environment by its user's original customization. Some of the well-known examples of a "customizing type" are wood materials with good workability and furniture with metal materials. In addition, IT devices such as smartphones and car navigators are also an example of a customizing type which are all over the market. In a new field, there is an example in which the product customizes the light level to the environment or the user's taste by using a light control glass. Furthermore, several customizing type value growth designs are appearing in nursing and medical fields, such as products that can optimize its shape to an individual (e.g. nursing care spoon with shape-memorizing material). Therefore, these fields are expected for future development.

Figure 4.3 : 4 types of value growth artifacts (3)

・学習型：

　使用に伴い，学習しながら成長するタイプ．「**学習型**」については，まず，PARO，AIBO などのロボットを思い出す．特に，PARO は近年，高齢者にとって，あたかもペットのように無くてはならない，かけがいのない存在になっており，愛着などの精神価値における価値成長デザインとして注目されている．また，ロボット以外では，たとえば，圧電素子や強誘電体などのセンサやアクチュエータが挙げられ，自動車をはじめとしてさまざまな制御機器類に利用されている．この領域は，今後のインテリジェント材料やスマート構造物として，さまざまな学習型の製品として利用されていくことが考えられる．

　なお，以上の4タイプの価値成長の実現のためには，ユーザーの好みに合わせたカスタマイズや製品の自然変化や馴染みを促進させるための「**サービス技術**」や，製品に生命が持つ学習機能，記憶，遺伝などのシステムを組み込む「**バイオ・インスパイヤード技術**」が有効であり，すでにいくつかの取り組みが進んでいる．

学習型

蓄積された情報に基づき，学習を繰り返すことで，ユーザに適合する

● PARO

図4.4：価値成長の4タイプ (4)

• Learning Type:

"**Learning type**" is a type of value growth design, in which the product adapts to its user by learning repeatedly based on accumulated information. The most representative example of a "learning type" are robot pets such as PARO and AIBO. Especially, PARO has become a one and only pet to the users (elders), and is getting attention as a value growth design with mental value (attachment). Other than robots, sensors and actuators for piezoelectric element or the ferroelectric substance used in vehicles and other control instruments are also an example of the learning type. It is thought that this field of intelligent materials and smart structure can be applied to products with learning type value growth design.

In order to realize the stated 4 types of value growth design the "**service technology**", which can consider customization of the product to the user and promotes natural changing and accustoming; and the "**bio-inspired technology**", which can input the life form's system of learning, memorizing, and inheriting to the product, is thought effective and some studies have already begun.

Figure 4.4 : 4 types of value growth artifacts (4)

価値成長デザインが拓く
新たなモノづくり産業と社会

　価値成長デザインがもたらす効果はさまざまである.

　モノづくり面では，つかうにつれて個別のユーザーや使用環境に適合することで，機能や特性の個別最適化が可能になる．また，時間経過に伴う価値観の変動にも対応することから，長期使用が可能な製品の実現も可能になる．これらの特長は，ユーザーが特定領域のプロフェッショナルであり，使い込んでいくタイプの製品には特に有利に働くと考えられている．そのため，楽器，医療機器，測定機器，生産財などには特に有効であり，今後，そのような製品が現れることを待望する．

　産業面から考えると，**モノづくり**（製造業，2次産業）と**コトづくり**（サービス業，3次産業）が結合した新ビジネスモデルの展開が可能である．これは，2次×3次産業の，もう一つの**6次産業**であるともいえる．この6次産業は，「おもてなしの心」など人々の精神的遺産を生かすこともできることから，固有の文化が反映される独自性を有する産業としても期待できる．

　社会面においては，製品の長期間使用による使い捨て社会からの脱却にもつながる．これは，使い捨ての社会から脱却し，製品を大切につかい，愛着を深める新たな精神価値社会の実現と相まって，資源エネルギー問題への対応や環境負荷の低減も自然と進むことが期待できる．このことは，従来まで行ってきた資源エネルギー問題や環境問題への直接的な対応策とは異なり，結果的にそれらの問題に対してサステナブルな社会を創ることにつながり，まさに，「北風と太陽」のおはなしにおける「太

New Product Creation Industry and Society Led by Value Growth Design

The effects that value growth design brings varies. The plant of value growth design to product creation enables the functions and characteristics of the product to optimize individually, by adapting to the user and use environment through usage. In addition, the value growth design can also adapt to the change of sense of values over time, enabling the realization of product with long-term usage. These traits of the value growth design are expected to be profitable in well-used products with professional of specific fields as users. Therefore, the plant of value growth design is especially effective for musical instrument, medical instrument, measuring instrument, and industrial materials. Thus, the appearance of products with value growth design are eagerly awaited.

From the industrial point of view, the combination of the **product creation** (manufacturing industry, secondary sector of industry) and the **value creation** (service industry, tertiary sector of industry) to create a new business model is expected to be possible. This new industry can also be said as the **6th industry** from the combination of the secondary and the tertiary sector of industries (2×3 industry). Since this 6th industry can make use of the mental inheritance of the people including "the heart of omotenashi", it is expected to be a unique industry that can reflect cultures.

From the social point of view, the plant of value growth design in product creation can lead to the breakaway from the

陽」的問題解決策であるといえる.

　以上に述べた価値成長デザインは,モノづくり産業や社会に新たな展開を創出する可能性を有している.そのため,今後,さまざまな領域での議論が行われることを期待する.

throw-away society by the long-term use of products. By breaking away from the throw-away society and becoming a society which uses products with care, can realize a new mental value society that deepens attachment the user feels toward the product. Then again, the new mental value society is expected to be able to correspond to the natural resource problems and reduction of environment load. As a result, the value growth design is connected in making a sustainable society, and to becoming a resolution for natural resource and environmental problems, like "the sun" from the story, "The North Wind and the Sun".

The value growth stated above, has a possibility to create a new product creation industry and society. Therefore, it is hoped that the discussion on how to plant value growth design to product creation in various fields is carried out in the future.

第4の知恵

「タイムアクシスデザイン」で拓く

独自の産業化と経済再生に向けて

4th Guideline

Adapting "Timeaxis Design" to Development

Toward Original Industrialization and Economic Revitalization

新製品のデザインを審査することがしばしばあるが，その際，私には常に気になっていることがある——この製品は果たして，5年後，10年後にどのような評価を受けているのであろうか．将来，業界における位置づけはどうなっているだろうか．社会にどのような影響を及ぼすにいたるのだろうか．

　元来，製品デザインの評価は，それが何年も使用された結果として行われるべきことであろう．つまり，製品の発表後，5年，10年を経過した後に，その使用の歴史を振り返ることで評価されるべきことである．しかしながら，多くのデザイン審査は製品の発表直後に行われる．そのため，我々審査員には，その後の社会の動向や未来の評価を予見しつつ，審査することが望まれる．このことは，審査員にとって重責であるとともに，難しい問題でもある．

　鍵を握るのは「**タイムアクシス**（時間軸）」である．「タイムアクシス」において，使用環境や製品の使われ方の変化，製品が使用される間の社会情勢や価値観の変動が，この製品の評価にどう影響を及ぼすのか．これらのことを想定し，総合的な評価を行うことが問われているのである．

　しかし，最近，ここである一つの考え方が注目されはじめている．それは，使用環境や使われ方などの「タイムアクシス」における変化に製品がいかに適合しつづけるかではなく，むしろ，製品自身がいかなる「タイムアクシス」の変化を創っていくかという新たなデザインの在り方である．以下に，その観点から「タイムアクシス」を考えてみたい．

I often judge new products in design contests. However, during the judging process, there is a moment I question myself, "What kind of evaluation will this product receive five years or decades later?", "How will this product positioned in the future industry?", and "What kind of influence will the product have on the society?".

Essentially, an evaluation of a product design should be given toward the result of the product's usage over the years. In other words, the product should be evaluated by looking back on years of history of its usage, and not just the moment it was announced. However, many design evaluation is carried out just after the announcement of a new product. Therefore, it is necessary for us, judges, to evaluate the product by foreseeing the future and social trends. This is a heavy responsibility and at the same time a difficult problem for the judges.

The "**timeaxis**" holds the key in solving the problem stated above. Through the perspective of "timeaxis", one must put the following questions into consideration when evaluating the product; how the environment and the usage of the product changes, and how the society and the sense of values changes affect the evaluation of the product.

However, there has been a new way of thinking that has begun to attract attentions recently. The new way of thinking of design is to understand how the product itself can create the changes under various "timeaxis" and not the adaptations of the products to the environment and usage changes it undergoes through "timeaxis". I would like to discuss about "timeaxis" from this perspective.

手工芸品に見られる価値成長デザイン

今からもう 25 年以上も前のことになる．当時，自動車メーカに勤めていた私は，自らの結婚式で次のような新郎の挨拶をした．「私たちが開発している車は，購入した時点が最も価値が高いようです．そして，その後，使用時間の経過とともに，残念ながら価値は徐々に下がっていきます．しかし，手工芸品には時間の経過とともに価値が上がっていくものも多くあります．私たちの結婚生活は，その手工芸品のように，時間が経つにつれて 2 人の絆を深め，価値を高めていきたいと思います」．今思えば，随分気恥かしい挨拶をしたものである．

ただ，ここで述べたいのは，私には当時から一つの残念な思いがあったということである．それは，自分たちがデザインしている自動車のほとんどは使うにつれて価値が減少してしまうこと．一方，漆工，革細工などの手工芸品に目を向けると，使えば使うほど価値が高まるものも多く存在する．このようなデ

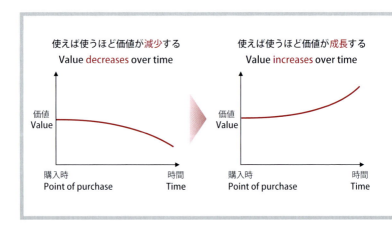

Value Growth Design Seen in Handicraft Work

More than 25 years ago while I was still working at an automobile company, I made a speech in my own wedding, saying, "Automobiles that we make nowadays have high value at the beginning (at the moment it is purchased). Unfortunately, automobile's value gradually decrease as time goes by. However, handicraft works have a characteristic in which its value increases as time goes by. I wish for our marriage to be like a handicraft work, in which the value grows over time by strengthening our bonds".

What I want to state here is that there is one thing that kept me feel disappointed from those days. The thing that made me disappointed is the fact that the value of the automobile we design decreases over time. On the other hand, many handicrafts work such as lacquer ware and leather work have a trait in which its value increases through usage and over time. From

図5：価値成長モデル

Figure 5 : Value growth model

ザイン（以下，「**価値成長デザイン**」と呼ぶ）を工業製品においても実現できないものだろうか？　もしそれが可能なら，使い捨ての社会からものを長く大切に使う社会へ転換できるかもしれない．私は当時，漠然とではあるがそのように考えていた．しかしながら，そのような価値成長デザインの自動車づくりは，私が自動車メーカに勤めた 14 年間，ついに実現しなかった．

　ところが，近年，科学技術の発展と相まって，ようやくそのヒントが見つかった．それは，時間軸に注目する「**タイムアクシスデザイン**」である．

新パラダイム「タイムアクシスデザイン」

　「**タイムアクシスデザイン**」は，デザインの理論・方法論に時間軸を導入するデザインの新たなパラダイムである．これまで，最適化理論やシステム工学などデザインに関わる学術領域においても，時間軸を積極的に取り入れてこなかった．

　タイムアクシスデザインでは，さまざまな時間軸のモデルを使用する．やや専門的な表現になってしまうが，時間軸の状態変化を示す「**非定常モデル**」，いったん状態変化を起こすと元に戻れない「**可塑モデル**」，秒・分スケールから時間・日・年スケールといった多層的なタイムスケールを同時に取り扱う「**マルチタイムスケールモデル**」など．これらのモデルにより，これまで難しかった多様な使用環境の変化や時間軸の価値変動などへの対応を目指している．さらには，想定外の問題への対応策

this fact, I had a vague question, Is there a way in applying the design with such trait (named as "**value growth design**") to the current industrial product? If the application of value growth design is possible to industrial products, it may help change the throw-away society into becoming a society, where people use products with care and for a long span. Still, even though I had such thought, I had not been able to realize such value growth design in my 14 years of engagement in automobile development.

However, I have found a clue from the recent technological development, which is the "**timeaxis design**", a concept of adapting the timeaxis to design.

New Paradigm "Timeaxis Design"

The "**timeaxis design**" is a new paradigm of design, which introduces timeaxis into design theory and methodology. However, there has not yet been such perspective in optimization theory and system engineering so far.

In timeaxis design, several models of timeaxis are used. The specialized terms of the models are; "**non-steady model**" indicating the change of state of a timeaxis, "**plasticity model**" considering the inability of restoring the state change after it has occurred, and a "**multi-timescale model**" expressing the multi-layer like time scale (such as seconds, minutes, hours, days, and years). These models are thought to correspond to the value changes from various environment and timeaxis, which was considered difficult until now. In addition, timeaxis design is also discussed as becoming a countermeasure for other

にまでも議論が及んでいる．

そして，これらのモデルを用いることで，価値成長デザインの実現に向けた研究開発が進められている．例えば，**価値発見期→価値実感期→価値成長期→価値定着期**といった時間軸上での価値成長を可能とする新たなモビリティシステム（車のような，ロボットのような製品）の開発が進められている．これには，犬猫のようなペットに対する時間軸の**価値成長メカニズム**を分析した知見が応用されている．さらには，そのモビリティシステムを他へ乗り換える際に，**価値伝承**を可能とする仕組みとサービスシステムを組み込んだ，まったく新たなビジネスモデルも検討されつつある．

「育つ」技術と「育てる」技術

タイムアクシスデザインを支える主な技術として，**バイオ・インスパイヤード技術**と**サービス技術**の二つが挙げられる．

バイオ・インスパイヤード技術は，製品に生命が持つ学習機能，記憶，遺伝などのシステムを組み込むことによる，製品が自ら「育つ」ための技術（**「育つ」技術**）である．これにより，生命の持つシステムを組み込まれた製品は，生命が有するロバスト性，冗長性，環境適応性などを有し，多様な使用環境においても安定した機能の維持や長期間の使用が可能になる．

サービス技術は，製品とその使用環境の両者にサービスを施すことによる，製品と使用環境の関係性を「**育てる**」技術である．この技術は，IBMが提唱した「**サービス科学**」およびその後に研究が進められている「**サービス工学**」で議論されている．サ

unexpected problems.

Research and development for the realization of value growth design has been promoted. For example, a development of a new mobility system (product like a robot and a vehicle) enabling the value growth design on several time axes: **value discovery phase**, **value realization phase**, **value growth phase**, and **value establishment phase**. The knowledge obtained from analyzing the **value growth mechanism** of the timeaxis for pets (cats and dogs) are applied to this system. Furthermore, a new business model that incorporated structure and service system enabling **value tradition** is discussed.

Technology "To Grow" and Technology "To Nurture"

The main technologies that sustain the timeaxis design are the **bio-inspired technology** and the **service technology**.

The bio-inspired technology is a **technology "to grow"** in which the product itself grows by incorporating the life form's characteristic such as learning, memorizing, and inheriting into a system. The incorporation of the life form's system into a product, enables the product to have the life form's trait such as robustness, redundancy, and environment adaptability. These traits allow the product to maintain stable function and provide long-term usage.

The service technology is a **technology "to nurture"** to provide a service to both the product and its usage environment to nurture their relationships. This technology is discussed in

ービス技術は，サービスによるユーザーの好みに合わせた製品のカスタマイズや製品の劣化に応じたメンテナンスはもちろんのこと，製品の使用環境やユーザーの価値観などの時間軸変動にも対応の可能性を有しており，製品の長期使用や製品に対するユーザーの愛着深化などを具現化する．

　この二つの技術により，時間の経過とともに製品の価値が成長し，長い間大切に使用可能なタイムアクシスデザインが可能になる．もちろん，一つの製品を長く使うと，製品は売れなくなり，経済面では不利になる．しかし，タイムアクシスデザインでは，モノづくり（第 2 次産業）とサービスビジネスによるコトづくり（第 3 次産業）との一体化を図り，サービスによるその不利な分以上のさらなる経済効果を上げることを狙っている．

　なお，この両技術は，いずれも日本人が得意とする技術ではないだろうか．前者のバイオ・インスパイヤード技術は，ロボット工学や制御技術における日本の先進性，後者のサービス技術は，日本人の細やかな「おもてなし」の精神に，それぞれ注目すれば理解できる．そして，これらの日本の独自性を生かせる技術をタイムアクシスデザインに用いることで，一つの新たな産業とその在り方を創生できる可能性を示唆しており，今後の展開が期待できる．

the "**service science**" that the IBM proposed and in the "**service engineering**". The service technology has a possibility to correspond to the timeaxis change of product's usage environment and to user values (such as being able to provide product customization for the user and provide degraded product maintenance), which can lead to the realization of the long usage of the product and user's attachment to the product.

These two technologies, enables the timeaxis design, in which the value of the products grow as time proceeds, leading to the user to use the product with care for a long time. Of course, the product will not sell and become a disadvantageous in economic aspect if it is long lasting. However, the aim of timeaxis design is to unify the product creation (the second industry) and value creation from the service business (the third industry), and to improve the economy despite the disadvantages given before, by providing service.

The stated technologies are thought to be the strength of a Japanese. The bio-inspired technology can be understood as Japan's innovative spirits in robot engineering and control technology fields; the service technology can be understood as Japan's sensible "omotenashi" spirit. The application of these Japan's unique technologies to timeaxis design is believed to lead to a creation of a new industry and its meaning, and is expected future development.

世界を先導する産業の構築と経済再生
── 人々の精神的な遺産を生かす

　経済再生のための材料は，人々がこれまで受け継いできた精神的な遺産のなかにある，と私は考える．そして，それらの精神的な遺産は，タイムアクシスデザインを実行する上での大きな力となる．

　例えば，「たおやかさ」．震災直後に見られた，被災地で公衆電話や給水車にきちんと列をなし静かに順番を待つ人々，信号が消えた交差点で道を譲り合うドライバー．そこには，やさしさのなかに秘められた人々の芯の強さ，しなやかな強靭（きょうじん）さも感じる．この「たおやかさ」は，製品を大切に使う精神に反映され，持続可能な社会を実現するうえでの精神基盤となるとともに，短期・中長期のタイムアクシスにおける復興活動のよりどころになるのではないだろうか．

　「**極めるマインド**」も，タイムアクシスデザインを後押しするだろう．人々の誠実な人柄には，修練に基礎を置く「極めるマインド」が内在していると言われている．この精神の存在は，手工芸品やこれまでの工業製品に対するものづくりの在り方から明らかである．この「極める」精神は，時間軸におけるリデザインをしっかりと継続させ，復興においても，社会の持続可能性においても有効に働くものと考える．

　このように，人々の精神的遺産は，タイムアクシスデザインに有効に働くだろう．我々は今こそこれらの遺産を生かし，タイムアクシスをデザインする文化に根づいた人々の価値観とそれに基づく方法論を獲得することで，世界を先導する新たな産業の構築と経済再生を目指す時期に来ているのではないだろうか．

Industrial Construction and Economic Revitalization to Lead the World
— To Make Use of Mental Inheritance of People

I think that the materials for economic revitalization lies within the psychological inheritance that humans passed on until now. This psychological inheritance will become a huge strength in carrying out timeaxis design.

For example, "taoyakasa" (like gracefulness) seen in people after the Japan's earthquake disaster. The people calmly stood in line to use the public telephone and to get water, and the drivers gave way to each other at the crossing where the lights were not working. From these actions, I felt the fortitude of people hidden in gentleness and tough yet flexible spirit. I believe that this "taoyakasa" is reflected in the mind to use products with care, and can become the mind base to realize a sustainable society. At the same time, this "taoyakasa" may lead to the timeaxis design (short term and long term) to becoming the base of the reconstruction activity.

The "**mind to pursue**" will promote the timeaxis design. It is said that inside faithful personality of people lies the "mind to pursue" which bases on discipline. This existence of mind is proven from the way of the handicraft works and conventional industrial products found in product creation until now. This "mind to pursue" enables the re-design in timeaxis continuously, and is considered effective in reviving and sustain the society.

Therefore, the psychological inheritance of people can be effective in realizing the timeaxis design. It may be time for us to make use of these inheritance and to aim for the construction of a new industry and revitalization of a new economy.

第 **5** の
知恵

「AGE 思考」
で思いつく

本当に，AI は人間の創造力を超えるか？

5th Guideline

Idea Generation Using "AGE Thinking"

Will AI Surpass Human Creativity?

AIの進化

　人工知能（Artificial Intelligence，以下"**AI**"）の進化が注目されている．AIという言葉は，1956年に開催されたダートマス会議ではじめて登場した．以来，AIは，1980年代の**エキスパートシステム**開発が盛んに行われた2度目のブームを経て，今日の第3次ブームに至っている．

　第3次ブームは，**多層構造**の**ニューラルネットワーク**を用いた**ディープラーニング**によるものである．2012年，「AIが猫の概念を理解した」というニュースが話題になった．スタンフォード大学とGoogleの研究グループによるものであり，映っている画像から，AIが猫の「**特徴抽出**」を行ったのである．この重要な点は，AIが事前に正解の画像を与えられていないにもかかわらず，猫という「特徴抽出」を自ら獲得したことにある．つまり，AIが教わってもいないある種の概念を発見しはじめたのである．

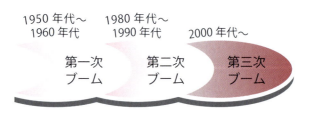

図6：AIブーム

Evolution of AI

Currently, the rapid evolution of the **Artificial Intelligence (AI)** is gathering attentions. The word AI first appeared at the 1956 conference held in Dartmouth College. From then, AI became a boom (2nd wave) in the 1980's, due to the growth of **expert system** development in the 1980's, and is presently becoming a boom (3rd wave) again.

The 3rd wave of the AI boom was caused by the **deep learning** of the **multilayered neural network**. In 2012, the news which introduced a research proposing "AI with an understanding of the image of a cat face" became popular. This research was about an AI performing a "**character extraction**" of the cat face from a picture, and was done by the Stanford University and Google. The important point in this research is the fact that the AI could do "character extraction" by itself without getting any correct solution beforehand. In other words, the AI could find and understand a kind of image without getting educated.

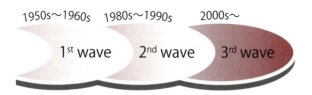

Figure 6 : AI waves

車を成長させる AI

　実は私たちも，2011 年に「未来モビリティ創生プロジェクト」（研究代表：松岡）のなかで，すでに「特徴抽出」に成功している．車に搭載した AI が自ら，道路環境やドライバー・車両の状態などの膨大な情報から走行場の特徴を抽出し，どのような走行環境なのかを的確に判断することを可能にした．さらに，その特徴を用いたシミュレーション走行実験の結果，変化する走行場を車が適切に認識し，その場に適した運転制御を継続的に行うことで，走れば走るほど燃費やバッテリー寿命が向上した．未来の「成長する車」開発の第一歩となった．

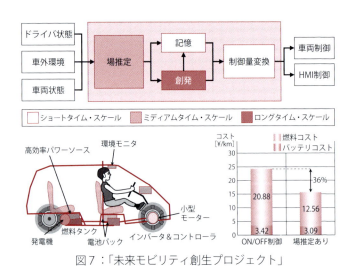

図 7：「未来モビリティ創生プロジェクト」

AI Nurturing Automobile

Actually, we have already succeeded in realizing "character extraction" in "Future Mobility Generation Project" (project director: Matsuoka) done in 2011. This project made possible for the on-board AI to extract the characteristics of the driving circumstances from various information (such as road, driver, and vehicle conditions), and understand the features of the driving circumstances precisely. In addition, in our running experiment by simulation, the "character extraction" of the AI toward the driving circumstances improved the fuel consumption and the battery life through driving, by perceiving the conditions changes and providing precise operation control. As a result, our project became the first step of developing "the vehicle that can grow".

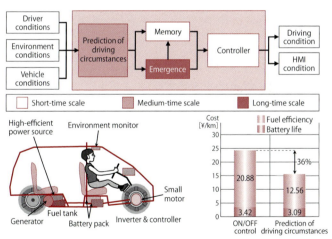

Figure 7 : "Future Mobility Generation Project"

今こそ，人間の創造的思考を問う

　AI は，他にも進化を示し，今や創造力を持ちはじめている．IBM のワトソンは，2011 年にクイズ番組で人間のクイズ王に勝利し，2015 年には料理のレシピ集を発表．さらに翌年には，白血病患者の病名を診断し，患者の命を救ったと報道されている．2016 年には，米グーグルの子会社 DeepMind が作成した囲碁対戦用 AI「AlphaGO」が人間のプロ囲碁棋士に勝利した．また同年，オランダの総合金融機関 ING グループらが，17 世紀の画家，レンブラントの題材や筆づかい，色合いなどの特徴をディープラーニングで分析し，それをもとに作成したレンブラントらしい図案を発表している．

　このように，近年，AI はさまざまな領域で，急激に創造力を身につけてはじめている．このような背景下，R. Kurzweil は，2045 年に AI の能力が人類全体の理解や発想の能力を超える「**シンギュラリティ**（技術的特異点）」が来るという．いわゆる「2045 年問題」である．

　これに関して，私自身はやや懐疑的であり，また，このようにいたずらに人類の不安をあおることを好ましくは思っていない．しかしながら，科学の目で見ると，このような AI の進化はやがてはやって来る可能性が十分ある．そのため，これを機に，人間の創造的思考とは何かを問い，それに基づき，今こそ人間と AI との創造のありようについて考える良い機会であると考える．

Human's Creative Thinking is Challenged Now

The AI is showing evolution in various fields and is beginning to have creativity. Watson from IBM has won against a human quiz champion in a quiz show in 2011, and also introduced a food recipe booklet in 2015. Furthermore, it was recently announced that Watson saved a patient's life by successfully diagnosing the name of the disease (which was leukemia). In 2016, the AI "AlphaGO", developed by the DeepMind company (a subsidiary company of Google), won against a human professional Igo player. In the same year, the Netherlands general financial institution, the ING group used the technology of deep learning to analyze the painting characteristics of the 17th century artist, Rembrandt, and achieved to create an artwork with Rembrandt's touch.

Lately, as you can see, the AI are drastically achieving creativity in various fields. Due to these backgrounds, Ray Kurzweil states that, in 2045 "**Singularity**" will come; a point in which the AI's ability will surpass the human being's mind and creativity.

Against this statement, I was slightly skeptical, and thought that it was not appropriate to bring uneasiness to human kind in such way. However, from the scientific perspective, it is true that there is a possibility that such AI evolution will come in the near future. Therefore, I think that this is a great opportunity to rethink what human creative mind is, and to think of a way in which the creativity of the humans and the AI can coexist.

AGE 思考モデル

　私は長年，デザインの実務と研究に従事した経験から，一つの創造的思考のかたちにたどり着いている．それは，「人間が発想する際，何らかの分析を必ず伴う」ということである．ここでいう分析とは，決してデータを用いた定量的分析のみならず，「ああなると，こうなるかな…」といった漠然とした主観的仮説も含んでいる．私は何十年もデザイン行為を観察してきたが，いまだ，分析がないまま発想した事例を見たことがない．そうした経験から得られた創造的思考モデルが「**AGE（エイジ）思考モデル**」である．

　AGE 思考モデル（図 8 参照）は，**分析**（Analysis），**発想**（Generation），**評価**（Evaluation）の三つの思考で構成され，それらの頭文字をとり，命名された．まず，課題や問題が発生すると，人はその現象や問題の原因について分析する．そうして，その分析をしながらアイデアを発想する．続いて，そのアイデアの妥当性や適合性などを評価する．その結果，アイデアがふさわしいと評価されれば，その問題の解（答え）となる．また，そうでない場合には再び分析または発想に戻り，アイデア発想を行う．人間の創造的思考は，この三つの思考ループを繰り返す．なお，やや専門的な話になるが，分析，発想，評価は，それぞれ**帰納**，**仮説形成**，**演繹**と呼ばれる各推論を利用することが一般的である．

AGE Thinking Model

I have come to an understanding of one model of creative thinking, based on my long-term experience in design and study. The understanding I reached is that, "when humans generate something, a kind of analysis is accompanied". The analysis stated here, does not only mean the analysis with quantitative data but also the analysis on vague hypothesis. I have observed design acts for many years, but I have not yet seen generation without undergoing analysis. The model derived from such experience is called the "**AGE thinking model**".

The AGE thinking model (figure 8) is composed of and named from **Analysis**, **Generation**, and **Evaluation**. First, when a problem occurs, the person will analyze the phenomenon and cause of the problem. The person will then generate idea by analyzing the phenomenon and then, evaluate its validation and compatibility. As a result, the generated idea will become the solution when it is evaluated desirable. If not desirable, the person will go back to the generation or the analysis, and again generate new idea. These three steps are repeatedly performed in human's creative thinking. In addition, each analysis, generation, and evaluation uses **induction**, **abduction**, and **deduction** for reasoning in general.

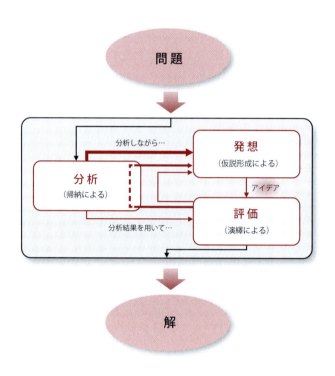

図8：AGE（エイジ）思考モデル

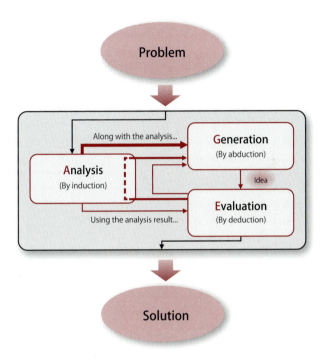

Figure 8 : AGE thinking model

AIは，人間よりアイデア発想が得意？

さて，皆さんはこの分析，発想，評価の三思考のうち，人間が最も得意なものは何だとお考えだろうか．一般によく耳にするのが発想である．発想こそが人間固有の強みであり，AIに勝るものであると……．

確かに，20年前には私もよくそのように考えていた．しかし，先述したように近年のAIの進化からすると，将来はむしろAIのほうが**アイデア発想**を得意とする可能性がある．そもそもアイデア発想は，**アナロジー（類推）型発想**と，さまざまな情報の組み合わせによる**統合型発想**の2通りの仮説形成によるものであり，いずれも情報の組み合わせによるものである．そして，考えうるさまざまな組み合わせを生み出すのは，コンピュータが得意である．しかも，**IoT**（Internet of Things）の情報収集能力からすると，AIが瞬時にアイデア発想する時代はそんなに先の話ではないだろう．将来は，AIのほうが発想力に勝る可能性があると言わざるをえないのである．

Is AI Better at Idea Generations Than Humans?

By the way, which one do you think human is best at, out of the three thinking: analysis, generation, and evaluation? Generally, most people will answer generation, and will say that the human being's ability to generate is superior to the AI's.

Indeed, I had the same opinion 20 years ago. However, as I stated above, there is a possibility that the AI will evolve to surpass the human ability of **idea generation**. This idea generation depends on two ways of abduction: the **analogical generation type** and the **integrative generation type** in which various information is combined. Both abductions are based on combination of information, which computers are relatively good at. Also, judging from the ability to collect information of **IoT** (Internet of Things), it may not be so far in the future where AI can instantly generate ideas. There is definitely a possibility for the AI to become superior at idea generation in the future.

意味づけは，人間の特殊な能力

　他方，分析と評価に関していえば，AIはいまだ多くの課題がある．分析については数値レベルの分析は可能だが，意味づけレベルまでの分析は苦手である．「猫の特徴抽出」はできるようになってきたが，「猫の何たるか」はまだ理解できていないのである．さらに，AIはいまだ**メタファー**（暗喩）やユーモアを理解できない．よく用いられるたとえだが，京都のお店で「ぶぶ漬けでもどないですか？」と言われても，AIは「そろそろ帰らなきゃ」と思うまでには，まだ時間を要するであろう．意味づけは人間の特殊な能力であるといえる．

　元来，意味づけは人間の特殊な能力であると私は考えている．意味づけのためには，**オントロジー**と呼ばれる知識の構造化などが必要であり，人間の脳はそれを備えている．その結果，占いや縁起担ぎなどをどうしても気にしてしまうのも，人間の特徴ではあるが……．

　評価についても，人間がいまだ優位であり，AIには課題がある．なぜなら，評価には感覚器などの身体が不可欠である．現在，ロボットやヴァーチャルリアリティなどの研究を通じて，さまざまな感覚器が開発されているが，職人や工芸家などのような高い感性を有するには，まだ時間が要するものと考える．

To Find Meaning is Special Ability of Human

On the other hand, AI still has several problems left when it comes to analysis and evaluation. In analysis, AI is strong at analyzing numerical values, but is still weak in analysis in finding meanings. The AI have become able to perform "character extraction of a cat", but still cannot understand "what a cat is". In addition, AI still cannot understand **metaphor** or humor yet. For example, people in Kyoto will say, "Would you like some Bubu-zuke (pickles)" which means, "We should be heading back now". However, it will take time for the AI to understand the meaning of the word. Therefore, it could be said that finding meanings is a special ability humans have.

I believe that the ability to create meanings is one of the human's specialties. To find meanings, **ontology** (structured knowledge) is necessary, and human brain is capable of structuring knowledge. As a result, humans tend to believe in fortune telling and omens.

Humans are also superior in evaluation, and AI still has some problem in evaluation. The reason to this is that, to evaluate, sensory organ or body is necessary. Currently, there are some studies using virtual reality (VR) and robots to create sensory organs, but it will still take time to realize high sensitivity of a craftsman.

AGE思考に基づく人間と
AIの共創システム

　このように，現時点では，AIの創造的思考にはまだ幾多の課題がある．しかし，今後，AIが進化を続け，創造力を身に付けていくことも事実であろう．そのため，今後のモノづくりの在り方としては，人間とAIがそれぞれ得意とする思考を見極め，両者のシナジーを考慮した**共創システム**の構築が有効であろう．

　AGE思考モデルが示す分析，発想，評価という各思考においては，人間とAIが発揮できる能力は，モノづくりの対象領域により異なる．しかも，時とともに変化する進化の度合いを考慮する必要がある．そのため，その共創システムにおける両者の役割分担は，AGE思考モデルの枠組みの中で常に時間軸で変化すべきであり，そのような**タイムアクシスストラテジー**の策定も必要になる．

　最後に，一言．AIは，所詮道具である．そのため，先の共創システムも人間中心でなければならない．そして，その視点に立って構築された共創システムこそが，真に有用で，持続可能なものとなると考える．

Co-Creation System of Human and AI Based on AGE Thinking Model

As you can see, creative thinking of an AI still has some problems currently. However, it is true that AI will gain creativity if it continues to evolve. Therefore, as for the way of future product creation, understanding which type of thinking the humans and the AI are good at, and constructing a **co-creation system** which considers the synergy of both, is thought effective.

In each analysis, generation, evaluation that the AGE thinking model shows, the ability AI can show depends in the target domain of the product creation. Besides, it is necessary to consider the degree of evaluation with the time change. Therefore, the role of the human and AI in the co-creation system must change with the timeaxis and within the AGE thinking model. Which means we must develop a **timeaxis strategy**.

Finally, I believe that AI is only just a tool. Therefore, the co-creation system in the future must be human centered. Then, the co-creation system created with this perspective will be truly useful and sustainable.

第6の知恵

「創発」に学ぶ

人はなぜ創造できるのか？

6th Guideline

Learning from "Emergence"

Why Can Humans Create?

人はなぜ創造できるのか？　この大問題を解き明かす鍵を握っているのが，「**創発**」である．創発とは，元来，自然現象を説明する言葉だが，近年，人のデザインという創造的な行為においても，同様の現象が存在することがわかってきた．この創造的なデザイン行為は，「**創めるマインド**」を持つものであり，従来は存在しない新たな発想を可能にする．

創発とは？——自然現象に学ぶ

　創発は，自然のシステムに存在する．それは，よく鳥の群れのたとえで説明される．

　群れの中の鳥は，近くの鳥や障害物との衝突を避ける．また，近くの鳥に速さと方向を合わせようとする．さらに，群れの中心に向かって飛ぼうとする．このような特性を単純なルールとして仮に設定した場合，これらのルールにより，近くの鳥どうしは互いに影響を与え合うことになる．そして，その影響が近くの鳥を通して次第に群れ全体に伝播し，群れの動きが決定される（図9参照）．

　この現象を言い換えると，個々の鳥がルールに基づいて動くことで，複数の鳥が「部分」的に相互作用（要素間の居所的な相互作用）し，やがてそれが群れ「全体」の動き（大域的な秩序）を創るということになる．このような「部分」から「全体」を創る特徴を「**ボトムアップ**」と呼び，創発現象においては要をなしている（図10参照）．

Why can humans create? The "**emergence**" holds the key to solve this great mystery. Emergence was originally a word used to explain natural phenomenon, but in recent years, it has been acknowledged that such emergent phenomenon exists in human's design process. This emergent design process holds the "**mind to originate**" and enables the generation of new ideas that did not exist before.

What is Emergence?
—Learning from Natural Phenomenon

Emergence exists in the system of nature. To describe this, the flock of birds is usually given as an example.

The birds inside a flock avoids nearby birds or obstacles. In addition, they adjust the speed and the direction of the flight to the other birds (flock). Furthermore, the birds try to fly toward the center of the flock. If such characteristic is established as a rule, by these rules, the birds nearby will affect each other. Then the affection is gradually transmitted through the nearby birds to the entire flock, resulting in the decision making of the movement of the flock (figure 9).

In other words, this phenomenon is expressed by the individual bird's movement based on the rule, several birds "partially" interacts with others (local interaction between elements), resulting into the emergence of the "whole" flock's movement (global order). Such characteristic to create the "whole" from the "partial" is called the "**bottom up**", and is the key in emergent phenomenon (figure 10).

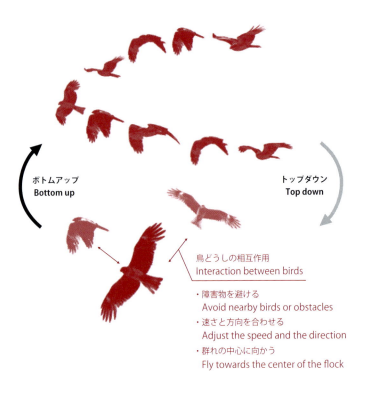

図9：鳥の群れにみられる創発現象

Figure 9 : Emergent phenomenon seen in bird flocks

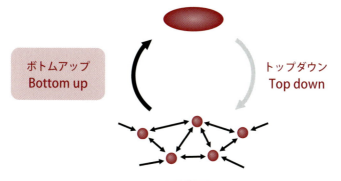

図 10：創発の概念

Figure 10 : Concept of emergence

二つのデザインの存在
──創発デザインと最適デザイン

　実は，優秀なデザイナーや設計者が，「創めるマインド」のもとに，何か新たな，独創的な製品をデザインする際にも，この創発と同じような行為をとることがわかってきた．そのような行為によりデザインすることを「**創発デザイン**」と呼ぶ．このデザインは，まず，何か気になる少ない製品特性や要件などの部分的な関係に注目し，そこから製品全体のシステムや機能を創出（発想）する**ボトムアップ型のデザイン**である．言い換えれば，新たな製品を生み出す創発デザインでは，初めからあらゆる製品特性や要件を考えずに，あえて特定の部分的な特性や要件に注目する．これは，改良を重ねることで最適化する「**最適デザイン**」とは異なるタイプのデザインである．

　最適デザインは，「極めるマインド」のもと，既存の製品を徹底的に調査し，その完成度を上げていく．このタイプのデザインは，たとえば既存製品のような何か前提となるものが存在し，それを基にして細部の製品特性や要件を分析・評価し，全体としての製品のシステムや機能を向上させていく**トップダウン型のデザイン**である．

　デザインには，この二つのタイプが存在し，両者はお互いに補い合う関係にある．一般に，何か新しい製品をデザインする場合には，まず「部分」から考える創発デザインを行う．それにより，「全体」としての新たなコンセプトや独創性なアイデアを創出する．そして，その後徐々に最適デザインに移行し，創発デザインで得られたアイデアの「全体」をもとに，「部分」である詳細を含めて，その完成度を上げていく（図11参照）．

Existence of 2 Designs
—Emergent Design and Optimum Design

As a fact, this emergence process happens when an excellent designer designs something new and original, based on the "mind to originate". Such designing process is called as "**emergent design**". This design is called the **bottom-up design**, which is a design that pays attention to the partial relations of the product characteristics or requirements at first, and then creates (idea) of the whole product's system and function. In other words, the emergent design pays attention not to the whole, but only to the partial characteristics and requirements of the product from the beginning. This design differs from the "**optimum design**", which optimizes by repeated improvement.

Optimum design thoroughly examines an existing product under the "mind to pursue" and raises its completeness. As for this type of design, called the **top-down design**, analysis and evaluates the product characteristics based on existing products, then improves the system of the whole product and improves its function.

In design, these two types of designs exist, and each have a complementary relation to other. Generally, emergent design is used to think from a "part" in order to design something new. Thereby this emergent design, generates new concepts and creative idea as a "whole". Then, the design gradually shifts toward the optimum design and accomplishes completion by refining the "part" based on the "whole" idea obtained by the emergent design (figure 11).

The precise interaction of these two designs is the key in

デザインの上流過程
Early process of design

概念デザイン　　　　　　　基本デザイン
Conceptual design　　　　Basic design

創発デザイン
Emergent design

全体
Whole

ボトムアップ　　　　　　トップダウン
Bottom up　　　　　　　Top down

部分
Partial

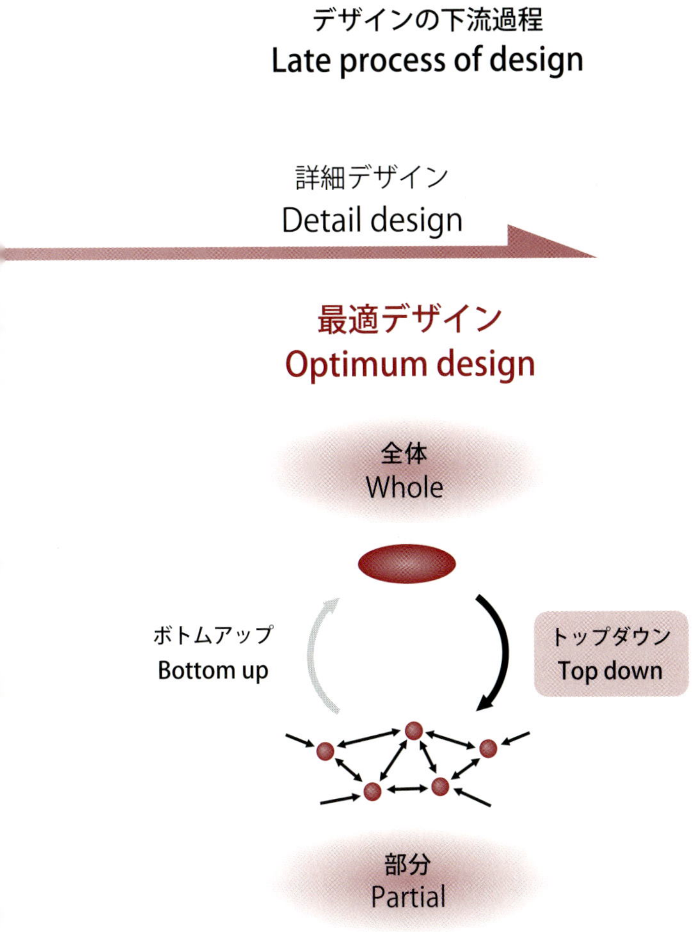

デザインの下流過程
Late process of design

詳細デザイン
Detail design

最適デザイン
Optimum design

全体
Whole

ボトムアップ
Bottom up

トップダウン
Top down

部分
Partial

図 11：創発デザインから最適デザインへ

Figure 11 : Emergent design to optimum design

この両者の的確な関わり方こそが，独創的でかつ完成度の高い製品を生み出す．そのため，近年，これに関するデザイン研究が進められ，その実態が徐々にではあるが，明らかになりつつある．

創発デザインの応用

　これまでに，「デザインには，創発デザインと最適デザインという二つのタイプが存在する」ことを述べた．しかしながら，このことについては，製品開発に携わっている方であれば，実は何となく感じとっているのではないだろうか．そして，そのような気づきが，すでにモノづくりの現場において，一つの重要な視点になっているものと考える．

　しかし，創発という自然現象に注目し，その気づきを具体的な両デザインの成り立ちとして示すことで，その視点を製品開発システムとして応用することが可能になる．ここに，創発デザインを論考する意義があると考える．

　先述したように，製品開発のプロセスは，創発デザインから最適デザインへと徐々に移行する．そして，その両者の的確な関わり方が，独創的でかつ完成度の高い製品を生み出す．そのため，デザイン研究において，その関わり方について，さまざまなシステムやノウハウが提言されており，企業の製品開発や大学のデザイン教育において，すでに応用されはじめている．興味のある方は，関連書籍や文献を参照されたい．

realizing a creative and high-quality product. Therefore, in recent years, study on these designs are proceeded, and its essence has been gradually clarified.

Application of Emergent Design

So far, "the existence of two types of design called the emergent design and the optimum design" has been stated. Then again, perhaps this fact has been unconsciously perceived to people associated in product development. It is thought that this kind of awareness is the most important perspective in the field of product development.

However, by learning from the natural phenomenon called the emergent, and specifically showing the scheme of both designs, emergent design enables the application of the perspective to product development system. Therefore, it is considered that there is a significance in learning emergent design.

As stated before, the process of product design is gradually shifting from optimum design to emergent design. As a consequence, precise relation of the two designs generates a more unique and completed product. Therefore, various systems and know-how about the relation of design study are proposed and applied in product development of the company and design education at universities.

創発デザインによるさらなる価値創造を

　これまでのモノづくりは，どちらかといえば最適デザインが得意であったと，よく耳にするのではないだろうか．確かに，「極めるマインド」の下に製品の改良を重ね，完成度を高めることは，これまでのモノづくりの得意技かもしれない．

　しかしながら，私見では，これからのモノづくりは「創めるマインド」の下に，イノベーティブな新製品を生み出すことが不可欠である．そして，世界を先導する，新たな価値創造が強く求められているのである．

　冒頭に述べたが，人はなぜ創造できるのか，この疑問に対して，以前のデザイン研究では的確に答えることができていなかった．しかし，創発への気づきが新たな展開を萌芽させ，創発の概念とそのメカニズムのつかみがデザイン研究を大きな飛躍へと誘導した．その結果，新たな製品を創出させてきたデザイナーや設計者の「デザインする」という不思議な行為について，その本質が少しずつ見えてきた．

　その本質さえ見極めることができれば，それぞれの企業の開発環境に合わせつつ，デザイナーや設計者のデザイン展開に有効な開発システムを構築することができるであろう．それにより，新たな価値創造に向けて，新製品の開発の新たな展開が期待できるのではないだろうか．創発デザインは，そのための大切なヒントを与えてくれるものと考える．

Further Value Creation Using Emergent Design

The manufacturing in Japan so far, has been assumed to be the optimum design. In fact, to improve products and raise completeness under the "mind to pursue" is one of the strength of Japan's manufacturing.

However, I believe that there are more to Japan's manufacturing. Under the "mind to originate", Japan has produced many innovative products. Moreover, product developments such as these, will be strongly expected from now on.

As stated in the beginning, the precise answer to "why humans can create?" has not yet been given in the former design study. However, the awareness of emergence led to an expansion, and the learning from the concept of emergence and its mechanism guided design science to a big leap. As a result, the essence of the interesting act called "designing", which designers undergo to create new products, has been gradually clarified.

If the essence of "designing" is cleared, can construct an effective development system which can adapt to individual company's development environments and design development of the designer. Thereby, the expansion of a new product development for new value creation can be expected. Thus, emergent design can be understood as an important hint to realize a new product development that can create new value.

第7の知恵

「状態」を考える

製品のデザインから関係性のデザインへ

7th Guideline

Considering "State"

From Product Design to Relationship Design

デザイナーとエンジニアのあいだ

　デザイナーとエンジニア（設計者）のあいだには，何があるのか？　そのようなことを，よく考えてしまう．確かに，デザイナーとエンジニアのあいだには，デザインの対象（製品特性），方法，利用する知識などにおいてさまざまな違いがある．そして，これらの違いが製品開発において補完関係となることで，有効に機能していることも周知である．このことは，すでにデザイン方法論の多くの議論において言及されている．しかしながら，さまざまなデザインを通じて一つ実感することがある．それは，両者におけるマインドには違いが存在し，その違いが製品デザインの質に影響を及ぼしているということである．

　概して，エンジニアは，「**極めるマインド**」を基本にしているように見受けられる．製品開発の際，まずエンジニアはユーザーや製品が使用される環境などの場をしっかりと把握する．なぜなら，製品の機能性は場に依存するためである．そして，その場を前提にし，それに適した新製品のデザインを行おうとする傾向が認められる．これを山登りにたとえるならば，一つの山に注目し，その山頂を極めようとすることに相当するのではないか．デザイン論でいえば，いわゆる「最適デザイン」型のマインドといえ，製品デザインの完成度に関与する．

　一方，デザイナーは，「**創めるマインド**」に重きを置いているのではないだろうか．一般に，デザイナーは，従来の製品とその場との関係性に注目する．そして，その製品と場の関係性に関して，従来とは異なる新たな関係性を創発しようとしているように見受けられる．つまり，新製品を市場に投入することで，

Between Designer and Engineer

What difference is there between a designer and an engineer? I often think of such a thing. In fact, there are many differences between a designer and an engineer such as, the design object (product properties), methods, and knowledge used. It is also well-known that these differences become the complementary relationship in product development, and leads to effective functionality. This fact has been already mentioned in many discussions of the design methodology. However, there is one thing that I realized through various designs. There is a difference in the mind of a designer and an engineer, and the difference affects the quality of the product design.

Generally, it seems that an engineer is based on the "**mind to pursue**". In product development, an engineer at first, understands the usage environments and circumstances of the user and product. The reason to this, is because the functionality of a product depends on the circumstances. Then, the circumstances become the prior conditions, and it is tended to carry out a new design of a product suitable for the given circumstances. This design process can be described as a mountain climb, which the climber only pays attention to one of the mountains and tries to attain the top of the mountain. It is called the mind of "optimum design" type in design theory, and plays a huge role in raising completeness of the product design.

On the other hand, a designer seems to be focusing on the "**mind to originate**". In general, a designer pays attention to the relationships of the conventional product and its circumstances.

場との新たな関係性を目指す立場である．これは，一つの山を極める前に，まず新たな山を目指すことに相当する．このことをデザイン論でいえば，やや難しい表現になるが，多峰性問題において新たな位相を目指す，いわゆる「創発デザイン」型のマインドに相当する．

デザイナーとエンジニアのあいだには，このマインドの違いが多々感じられる．決して，エンジニアが「創めるマインド」を持たず，デザイナーが「極めるマインド」を軽視しているわけではない．しかしながら，あくまで相対的な関係としてそのような傾向が見受けられ，そのマインドの強弱が，製品デザインの質に大きく関与していることは事実であろう．

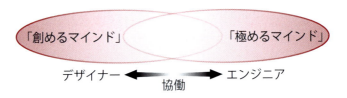

図12：「創めるマインド」と「極めるマインド」

Then, the designer tries to generate a new and non-conventional relationship of the product and its circumstances. In other words, the position of a designer is to aim to generate a new relationship with the circumstances, by bringing a new product to the market. This act could be described as a mountain climber, who aims for another mountain before climbing on top of one mountain. It is called the mind of "emergent design" type in design theory, which aims for a new phase in multimodality problem.

The differences of minds between a designer and an engineer could be frequently noticed. Nevertheless, it does not mean that the engineer thinks lightly of the "mind to originate" and the designer thinks lightly of the "mind to pursue". However, such traits of each minds are only given from the relative relations, and as a fact, it is the power of either minds that gives a huge influence to the quality of a product design.

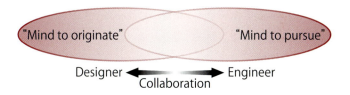

Figure 12 : "Mind to originate" and "mind to pursue"

場に適したデザインから，新たな場を創るデザインへ

　元来，この二つのマインドは，製品開発において相補関係にあり，双方が必要とされている．新規性を求めるデザインにおいては，まず「創めるマインド」により新たな製品と場との関係性を創発することが必要である．次に「極めるマインド」によりその新たな関係性の最適性を高めていく．この二つのデザイン過程を伴うか否かが，デザインの質を大きく左右することになる．

　一般に，改良型のデザインにおいては，「極めるマインド」により，最適デザインが行われる．改良型のデザインにおいては，既存の場を前提にすることで，製品の最適化を進めるためである．しかしながら，新規性を求めるデザインにおいては，先に述べたように，「創める」と「極める」の双方による新規性と最適性の共存が不可欠であり，その共存がデザインを行ううえで重要な鍵を握ることになる．

　新製品が市場に投入された際，新規性と最適性の共存が市場の評価を左右することが多い．特に，製品と場の関係性において新規性を創り出したデザインは高い評価を得たものと考える．このようなデザインは，従来の場を前提としていない．むしろ，新製品を市場に投入することにより，新たな製品と場の関係性をデザインしているといえる．さらに，このことは，結果的に新たな場を創ることにもつながる．言い換えれば，製品と場の新規な関係性をデザインするために，従来の場を再考することで，新たな場を創るという立場をとっていることになる．そして，このような立場をとるデザインこそが，その製品の新しい可能性を創造することになる．

From Design Suited to Circumstances to Design That Creates Circumstances

These two minds are complementary in product development, and both minds require each other. In a design which seeks novelty, one must at first, generate the relationship of the new product and its circumstances using the "mind to originate". Next, one must use the "mind to pursue" to optimize the new relationship. Whether one performs these two design processes or not, will greatly affect the quality of the design.

Generally, in an improvement type of design, optimum design is performed with the "mind to pursue". In an improvement type of design, the conventional circumstances are conditioned to promote the optimization of the product. However, in a novelty type of design, the coexistence of both the novelty and optimum, which are the mind to "originate" and "pursue" are necessary. The coexistence of these two minds are an important key in designing.

When a new product is brought into the market, it is up to the coexistence of the novelty and optimum which leads to the evaluation in the market. Especially, the design that created novelty in the relationship of the product and circumstances is thought to earned high evaluation. Such design is assuming non-conventional circumstances. Rather, it may be said that the bringing a new product to the market, is a designing of a new relationship of the product and its circumstances. Furthermore, this leads to the creation of a new circumstances. In other words, the act of reconsidering the conventional circumstances to design a new relationship of the product and circumstances, means to create new circumstances. Such design can generate the product's new possibilities.

製品と場の関係性を示す「状態」に注目

製品とそれが使用される場との新たな関係性をデザインするためには,その指標として「**状態**」を用いるのが有効である.「**状態**」とは,場に影響を受ける特性であり,色彩,材料,形状などの製品そのものの特性である「**属性**」とは異なる.

たとえば,製品の剛性や強度などに関する力学面では,材料のヤング率は「属性」であるが,荷重条件や拘束条件に影響を受ける応力分布は「状態」である.製品の剛性評価は,当然のことながら荷重条件などの場の影響を受ける.そのため,応力分布などの「状態」を評価指標としてデザインを進めることは必然である.

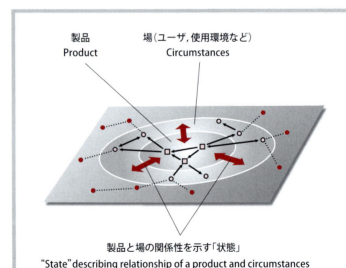

製品と場の関係性を示す「状態」
"State" describing relationship of a product and circumstances

Considering "State" Which Describes Relation of Product with its Circumstances

It is effective to use the "**state**" as an index to design a new relationship of the product and its use circumstances. "State" is a property affected by the circumstances, and is different from "**attribute**" properties such as the product's color, materials, and shape.

For example, the dynamic properties (e.g. stiffness and intensity) of the product's material, such as Young modulus is an "attribute", but the stress distribution affected by the load conditions and restriction conditions is a "state". As a matter of fact, the product's stiffness evaluation is affected by the circumstances, like the load conditions. Therefore, it is

図 13：状態

Figure 13 : State

このことは，造形面においても同様である．製品そのものの色彩や形状は「属性」であるが，光源の影響を受ける色味や陰影の見え方は「状態」にあたる．製品の造形性評価は，本来，場の影響を受ける．光源の演色性や自動車のハイライトなどを考えれば，容易に理解できるであろう．そのため，造形面においても，後者の「状態」に注目してデザインを進めることが肝要である．

　しかしながら，これまでの造形面に関するデザインにおいては，「属性」を直接評価する傾向が否めない．例えば，「○○な形状が△△な評価を受ける」などである．この傾向は過去の学術研究においても同様の傾向が認められる．ところが，「属性」の評価結果は，少しでも場が変化すると使えない場合が一般的である．そのため，今後は，造形面においても「状態」を用いるデザインを推進する必要があると考える．

　元来，デザインの本質は，製品をデザインするのではなく，機能性（造形性や操作性などを含む）をデザインすることであろう．そして，このことを言い換えれば，製品と場の関係性をデザインすることになり，ひいては「状態」に注目してデザインすることの重要性を意味することになる．

necessary to promote design with an "state" as an evaluation index.

This is similar to molding. The product's own color and shapes are an "attribute", but the color taste and shades of how the product looks is affected by the light, which is considered as a "state". The original molding evaluation of a product is affected by the circumstances. It may be easily understood, if you imagine a color rendering of light source or a vehicle's highlight. Therefore, it is also important in molding to pay attention to the "state" and undergo design.

However, the past molding design tends to directly evaluate the "attribute". For example, "a shape that is 'AA', receives 'BB' kind of evaluation". As for this tendency, a similar tendency is recognized in the past scientific studies. Nevertheless, the "attribute" evaluation result, in general, becomes unusable when the circumstances changes (even with the slightest change). Therefore, it is important to promote design that uses "state" in future molding.

The essence of design is not to design the product, but to design the functionality (includes molding characteristics and operability). In other words, this means to design the relationship of the product and circumstances, thus meaning in the importance of paying attention to the "state" when designing.

「状態」を用いたノウハウ伝承

　「状態」に注目してデザインを行うことは，**ノウハウの伝承**にも大きく貢献する．なぜなら，製品の機能性は場に依存するからである．場は多様であり，時とともに変化する．そのため，デザインのノウハウとして伝承すべきは，「属性」ではなく，製品と場との関係性を示す「状態」である．「状態」こそが製品と場との普遍的な関係性を示しており，デザインの知恵や工夫として伝承されるべきノウハウなのである．

　「状態」に注目してデザインすることは，その企業のデザイン力を強化することにつながる．製品デザインにおいては，単に新製品を開発するだけでなく，デザインのノウハウを構築する役割も担っているだろう．そのため，製品デザインを通じて得られた「状態」のノウハウをドキュメント化し，伝承することは企業にとってデザイン力を高めるための有効な手段となる．

　現在，他社に容易に真似できない製品開発力の強化が問われている．このような状況下，企業は独自のデザインノウハウを「状態」として構築し，伝承し，蓄積することが急務であろう．そして，その蓄積されたノウハウを用いることで，製品と場の新たな関係性をデザインすることが期待されているのではないだろうか．

The Knowhow Tradition Using "State"

The consideration of "state" in designing, can lead to a contribution of **knowhow tradition**. This is because, the product's functionality depends on its circumstances. Circumstances has many varieties and changes with time. Therefore, it is not the "attribute" but the "state", which shows the relationship of the product and its circumstances, is important in passing down the knowhow of the design. The "state" that shows the general relation of the product and its circumstances, and is the knowhow that should be passed down as wisdom and ingenuity of the design.

In addition, the consideration of "state" in designing, leads the strengthening of the company's design. Product design takes the role of developing new products and also to construct a knowhow of the design. Consequently, by creating a document on the knowhow of the "state" obtained through product design and passing it down, will become the effective means to raise the design abilities of the company.

Currently, improvement of product development technology that cannot be imitated by other companies is being required. Under such conditions, it is an urgent task for the companies to construct, pass on, and accumulate their own design knowhow as a "state". It is expected that a new relationship of a product and its circumstances can be designed by employing the accumulated knowhow.

第 **8** の知恵

「ロバスト性」を獲得する

強靭で，持続可能なものづくりと社会のために

8th Guideline

Ensuring "Robustness"

For Tough and Sustainable Manufacturing and Society

大震災の「想定外」に学ぶ

「**想定外**」——この言葉を，2011年3月11日の東日本大震災以降，私たちは何度耳にしたことだろうか．この言葉は，安全神話の崩壊の象徴としてマスコミに幾度となくつかわれ，それはあたかも，人々の科学技術に対する不信感や不安感をあおっているかのようにも見える．

ここで，想定外には二つの意味があることに注目する．その一つは，優先度の低さから「**想定しなかった想定外**」．もう一つは，予想もつかないレベルの「**想定できなかった想定外**」である．

前者の「想定しなかった想定外」は，現在の科学技術で予想することの可能な事象である．しかし，その発生確率や重要度の低さからコストなど他の要因が優先され，結果として開発段階における想定からは外されてしまったものである．このような想定外は，実際のところかなり多いのではないだろうか．今回の震災における津波対策や原発の安全対策の問題もこれに該当する．

国際原子力機関（IAEA）の関係者らは，今回の原発事故に対して「比較的コストのかからない改善をしていれば，完全に回避できた可能性がある」と指摘している．この指摘内容の是非とその効果は定かではない．しかしながら，その可能性は十分にありえ，この指摘は，さまざまな問題の重大性を正確に把握し，的確な総合的判断を行うことの重要性を改めて認識させてくれる．

一方，後者の「想定できなかった想定外」は，いくら科学技術の粋を尽くしても予想できない事象を指す．この場合，開発

Learning from "Unexpected" of Great East Japan Earthquake

How many times have we heard the word "**unexpected**" after the Great East Japan Earthquake struck us on March 11, 2011? The word "unexpected" appears in the media for several times as if it was the symbol of disruption of the safety myth of Japan, giving people uneasiness, which leads to the distrust toward technologies.

Here, we will pay attention to the two types of meanings "unexpected" has. First is the "**was not assumed**", due to the low priority of the phenomenon. Second is the "**was unable to be assumed**", due to the level of the phenomenon being far from expectation.

The "unexpected" from "was not assumed" is a phenomenon that could have been expected by current technologies. However, due to the phenomenon's low possibility of outbreak and priority, other factors are prioritized in using the budget for development. As a result, the phenomenon that could have been expected and prevented is excluded during the development stage, leading to the "unexpected". This kind of "unexpected" may be the most frequently occurred problem in development, and the tsunami disaster and nuclear melt down caused by the earthquake on March 11, 2011 falls under this category.

The International Atomic Energy Agency (IAEA) points out that, "The Fukushima Daiichi Nuclear Disaster could have been avoided if precise measures with decent cost were taken". The right or wrong toward this statement and its effects are not yet sure. However, there are possibilities that this indication may be

段階での対応は不可能である．しかし，仮に開発段階では想定が不可能であったとしても，その後の使用段階において科学技術が進展し，想定とその対応策が可能になる場合も考えられる．特に，原発のように長期に使用する際には，そのような可能性が高いのではないだろうか．ただし，一般に，使用段階における対応策の織り込みには多くの費用がかかる．そのため，この想定外においても，前者の「想定しなかった想定外」と同様に，その問題の重大性や発生頻度と他のさまざまな要因を総括した的確な判断が肝要であるといえよう．

　以上に示したように，「想定外」に対応するためには，いずれにしても的確な総合的判断が重要な鍵を握ることが理解できる．しかしながら，実際の開発においては，「あちらを立てれば，こちらが立たず」といった**トレードオフ問題**（二律背反問題）が多いのが実情であろう．そして，このトレードオフ問題こそが，「想定外」の根源なのである．

　しかし，このようなトレードオフ問題に対して，近年のものづくりはただ手をこまぬいているわけではない．一見トレードオフに見える関係も，工夫次第では両立が可能となることも多い．そして，今，その方法論として注目されているのが「**ロバストデザイン**」である．

effective, and lets one recognize the seriousness of various problems and importance of giving precise and general judgment.

On the other hand, the "unexpected" from "was unable to be assumed" is a phenomenon that cannot be expected even with current technologies, and is impossible to make countermeasures during development stage. However, even if the assumption is impossible at the development stage, as technology progresses, the assumption and countermeasures can be put into consideration. This technological development can be said especially in nuclear power plant with long-term usage. However, the costs are likely to be very expensive when inserting countermeasures in use stage. Therefore, it is also important to give precise and general judgment by understanding the seriousness of the problem, possibility of outbreak, and various factors.

As stated above, in order to prevent "unexpected", the ability to be able to give precise and general judgment can be understood as the key factor. However, in actual development exists many **trade-off problems** (antinomy problems) like, "It is hard to please all". Then again, this trade-off problem is the root cause of the "unexpected".

Nonetheless, present technologies are not just standing and watching against this trade-off problem. There are relations that may seem like a trade-off problem, but depending on how the problem is managed, can enable the coexistence of the relations. Thus, methodology called the "**robust design**", is currently drawing attentions.

多様な場に対応する「ロバストデザイン」に注目

「**ロバスト**」とは，「強靭な」などと訳される．ただし，その言葉には，多様な場（外部環境，使われ方，条件など）に対して常に安定した強靭さを有するという意味が込められている．そして，そのロバスト性を確保するために，考えられうるさまざまな手段のなかから最適な手段を選定し，製品化する方法論が「**ロバストデザイン**」である．

たとえば，強靭であるためには，堅固な強靭さが常に望ましいとは限らない．風にしなる竹が持つしなやかな強靭さが適切な場合もある．強靭さを持たせるための手段はさまざまあるのである．そして，それらの手段のうち，多様な場に対して安定的に機能を確保するためにロバストデザインが必要とされている．

元来，ロバストデザインは，もののばらつき（寸法誤差，材料のばらつきなど）に対して，常に安定した機能を確保するための方法論としてスタートした．田口玄一博士が提唱したいわゆるタグチメソッドは有名であるが，近年では，世界中でさまざまな方法が開発され，急成長している．たとえば，起こりうる多様な場とその発生頻度や重大性を考慮することで，常に安定した機能を確保する手法が提案されている．また，開発段階で，多様な場に対応するために調整機構が必要か否かを検討することがあるが，その必要性に関する判断やその機構の調整範囲を最も効果的にかつ最小にするための方法なども開発されている．さらには，製品群のバリエーションやラインアップをいかに設定すれば，市場の多様なニーズを効率的にカバーできるかといった企画の問題にも対応可能なロバストデザインも研究されは

Paying Attention to "Robust Design" Corresponding to Various Circumstances

Robustness is translated as "toughness". However, robustness means to have stable toughness to various circumstances (external environments, surroundings, usages, and conditions). The "**robust design**" is a methodology in which the manufacture is done by choosing out the most suitable mean from various means to ensure robustness.

For example, toughness may not be always desirable to be stiff toughness. The flexible toughness that bamboo has may be appropriate at certain times. Thus, there are various ways to keep toughness. Out of all the techniques to keep toughness, the "robust design" is a methodology that can ensure stability of the functions for various circumstances.

The origin of robust design started from a methodology to ensure stability of the function against variability of a product (such as dimension error and unevenness of materials). To this day, the Taguchi method, developed by Genichi Taguchi is known as one of the most common robust design method. However, various methods are being developed and rapidly growing for the past years. For example, a technique to ensure stable function by considering the outbreak frequency and seriousness of various possible circumstances is being developed. In addition, examining whether the adjustment mechanism is necessary or not is important at a development stage. Nevertheless, the method to minimize the judgment toward the necessity of the mechanism and the adjustment range, to be the most effective are currently developed. Furthermore, the robust design, which can cover variety of

じめている.

　これらの新たなロバストデザインの登場により，従来では難しかったロバスト性の確保が可能となってきている．そのため，これらの方法をもっと積極的に活用することで，従来，対応が難しかったトレードオフ問題とそこから発生する想定外の問題を解決していく必要があるだろう．

製品の大規模化・複雑化，その功罪

　私には，デザイン審査をする機会がある．その際に一つ気になっていることがある．それは，審査対象である製品・人工物の多くが訴求点として「高性能化」や「多機能化」を挙げており，その結果，製品が大規模化・複雑化の一途をたどっている点である．確かに，高性能化や多機能化は製品価値を高めるうえで重要な開発の方向性であり，これらにより，高い生産性や利便性を実現してきたといえる．

　しかしながら，これらに伴う製品・人工物の大規模化・複雑化は，安心・安全の問題という深刻な副作用を内在させているのも事実であろう．大規模で複雑な製品・人工物は，概して，全体制御や詳細管理が難しい．そのため，これらの大規模化・複雑化は「想定外」を誘発し，安全を害する危険性を高める傾向は否めない．しかも，製品や人工物の大規模化・複雑化は，発生する被害の規模も大きくするという問題を内在させている．東日本大震災に伴う原発事故も同様であった．大規模化・複雑化した製品や人工物は，時としてその脆さを見せ，暴走をはじめるのである．

　このような製品開発での状況下，多様な場に対してロバスト

needs of the market effectively by setting the precise variation and lineup of the product, is also being studied.

The appearance of these new robust designs allows the securing of the robustness that was considered difficult before. Therefore, it is necessary to solve an unexpected problem caused from trade-off matter by employing the robust design methods.

Merits and Demerits of Large Scale and Complex Products

I have many opportunities to evaluate designs, throughout the evaluation process, I have one thing that concerns me. My concern is that the requirement of a product and an artifact targeted for evaluation are "advance technology" and "multiple functions". As a result, the society is leading toward large scale and complex product manufacturing. It is indeed, important to develop product with advanced technologies and multiple functions to raise its value, and it is a fact that these requirements have realized high productivity and convenience.

However, as a fact, the enlargement and complication of the product and artifact has let a serious side effect relating to the safety and reliability problem. Generally, the detailed management of a large scale and complex product and artifact is considered difficult. Therefore, the enlargement and complication causes the "unexpected", and it tends to raise the risk of violating safeness. In addition, the large scale and complex product and artifact, leads to the enlargement of the disaster. The Fukushima Daiichi Nuclear Disaster caused by the Great East Japan Earthquake can be given as an example of disaster caused from large scale and complex artifact. The large scale and complex

性を確保することは，ますます重要課題となっている．ロバスト性を獲得し，さまざまな使用環境下で安定した機能と安全性を有する製品・人工物の開発を推進し，ひいては，強靭で持続可能なものづくりとすることが，今，望まれている．

ロバストなものづくり産業と社会
── 世界一安心・安全な日本再生に向けて

　2011年の東日本大震災により，我々日本人は多くのことを感じとり，学んだのではないだろうか．そして，一説では，社会に対する価値観さえも変容しつつあるともいわれている．人との絆や優しさを重んじるとともに，ものに対しても使い捨てから，ものを大切に使う価値観に移行しつつある傾向が伺える．

　このような状況下，ものづくりにおいても，従来の経済効率性に視座を置く短期的最適性重視の傾向から，長期的な持続可能性に視座を置くロバスト性重視の価値観へ移行すべき時期に来ているのかもしれない．そして，そのようなものづくりを通じて，ものづくり産業自体にも，強靭で持続可能なあり方が問われているように考える．

　今こそ，我々はロバスト性に注目するものづくりを再認識し，それを実現することで，ものづくりの哲学と技術を獲得し，世界一安心で安全な社会を目指すべきではないだろうか．

product and artifacts are becoming disrupted due to its fragility.

To ensure robustness that can consider various circumstances under large scale and complex product manufacturing is becoming an important issue. Therefore, to promote the development of products and artifacts with reliability and functions that can be provided stably under various environment; and the realization of a tough and sustainable manufacturing is currently demanded.

Robust Manufacturing Industry and Society —Toward Reconstruction of the World's Most Reliable and Safest Japan

The disaster caused by the Great East Japan Earthquake in 2011 has given the people of Japan, a chance to think back and learn about the mistakes of current way of manufacturing. Also, from another point of view, the sense of values toward the society is changing. The respect toward the bonds and gentleness between people has shifted the sense of values from throwing away products to keeping products with care.

Under such conditions, manufacturing may be at the time when its senses of values are shifting from short-term optimality to establish aspect in the economical effectiveness, into long-term sustainability to establish aspect in the robustness. Through such manufacturing, the manufacturing industry is challenged to establish a tough and sustainable way.

I believe that we must aim to become the world's most safe and reliable society by acknowledging and realizing the manufacturing that considers robustness to achieve philosophy and technique of manufacturing.

第 9 の知恵

「多空間デザインモデル」で開発する

Ｍメソッド：新たな価値創造に向けて

9th Guideline

Applying "Multispace Design Model" to Innovation

M Method: Toward New Value Creation

「豊かさ」と「安心」のはざま
――大規模化・複雑化の功罪

　今日までのデザインは，人々の生活を豊かにしてきた．美しさ，機能性，利便性など人々のさまざまな「豊かさ」を満たすべく，多くの特性を製品に埋め込み，便利で快適な社会を実現してきた．そして，その実現のために，デザインは，製品の高性能化・多機能化を図り，その手段として構造やシステムの大規模化・複雑化を推し進めてきた．

　しかし，そこには，深刻な副作用の可能性が内在している．それは，安全性，信頼性，耐環境性などといった「安心」に対する問題の発生である．今日までのデザインは，自らの子である製品を大規模化・複雑化させて，育ててきた．その結果，肥大化させてしまった我が子を，時として制御できなくなっている．福島の原発事故や飛行機の故障，さまざまな環境汚染問題もその一例であろう．肥大化した製品は，時としてその脆さを見せ，暴走をはじめている．現在のデザインは，「豊かさ」と「安心」のはざまで苦悩しているといえるのではないだろうか．

Between "Abundance" and "Ease" —Merits and Demerits of Large Scale and Complexity

The design to date, has made our living affluent. To meet people's various "abundance" such as, beauty, functionality, and convenience, several characteristics were imputed into products to realize a convenient and comfortable society. To realize such society, design with advanced technology and various functionality are applied to products. As for this method, the large scale and complexity of the structure and system has been proposed.

However, serious side effect lies within this "abundance". The side effects are the outbreak of problems for "ease" such as safety, reliability, and environmental resistance. The design until today, has let the product grow larger and become more complex. As a result, the overgrown products have become difficult to control. The nuclear melt down in Fukushima and failure in aircrafts are one of those examples. Such overgrown products are gradually showing its fragility and are beginning to go out of control. Therefore, it can be said that the present-day design is agonized between the "abundance" and "ease".

デザインサイエンスの知恵を生かせ
——多空間のデザイン思考

　このような製品の大規模化・複雑化が生み出す問題を含め，現在のさまざまなデザインにおける問題を解決するために，**デザインサイエンス**が研究されている．デザインサイエンスとは，「デザインという人の創造的行為における法則性の解明と，デザイン行為に用いられるさまざまな知識の体系化」を狙いとした新しい学問である．現在では，新しいアイデアを生むためのデザイン思考のメカニズムや，斬新でかつ完成度の高い製品開発のための方法論など，幾多の研究成果が得られている．なお，デザインサイエンスに関する主要な用語に関しては「デザイン科学辞典」として，現在，インターネット上にて，**デザイン塾**のホームページ（http://www.designjuku.jp）において公開されているので，参照されたい．

　デザインサイエンスにおける研究成果の一つとして，「**多空間デザインモデル（Mモデル）**」が挙げられる．これは，あらゆるデザイン行為を一般性のあるモデルとして表現したものであり，デザイン理論の枠組みの一つとしても利用されている（図14参照）．このモデルは，**デザイン思考**とその原動力となる**デザイン知識**の二つで構成される．また，デザイン思考においては，考慮すべき膨大なデザイン要素を，**価値**，**意味**，**状態**，および**属性**といった多数の空間（多空間）に区分している点を特徴としており，特に大規模で複雑な構造やシステムを有する製品のデザインにおいては，この多空間の特徴が有効であることを，このモデルは説いている．

Making Use of Knowledge from Design Science — Design Thinking of Multispace

The **design science** is studied to solve various design problems including problems formed from producing large scale and complex products. design science is "an academic discipline focused on the clarification of the laws that govern deign as a human act of creativity as well as the systemization of the knowledge used in designing". Many research results on the mechanism of design thinking in generating new ideas and methodology in developing unique and high-quality products have been achieved till now. For further definition on design science terms, please visit the online "Design science dictionary" at **DesignJuke** website. (http://www.designjuku.jp)

One of the achievements in design science is the "**multispace design model** (**M model**)". This model generally expresses the design act, and has been used as a framework for design theory (figure 14). This model is composed of **design thinking** and **design knowledge** used in that design thinking. The design thinking explains the massive design elements by dividing them into several spaces (multispace) which are **value**, **meaning**, **state**, and **attribute**. This model preaches the effectiveness of the characteristic of the multispace, especially in designing products with large scale and complex system.

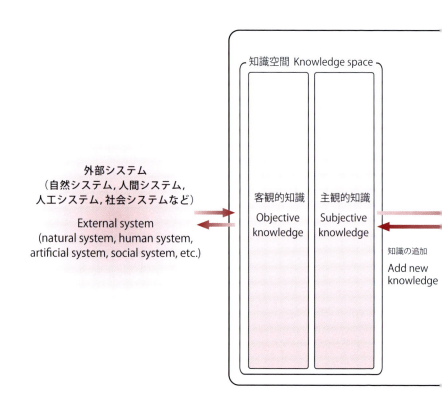

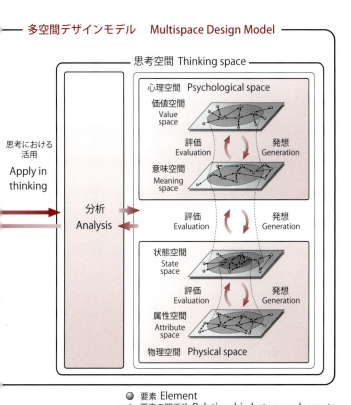

Figure 14 : Multispace Design Model (M model)

「M メソッド」
──「自由な思考」と「理にかなった思考」の両立

　「多空間デザインモデル」に基づいたデザイン方法として，「**M メソッド（多空間デザイン法）**」が提案されている（図 15 参照）．この「Mメソッド」は，デザインのみならず，経営，企画，研究，さらには一般の生活においても利用可能な思考メソッドである（図 16 参照）．この特長は，頭のなかで，「多空間を視点として，自分の好きなやり方で，分析しながら発想を行う」点にある．これにより，大規模な構造やシステムの製品においても，従来では難しかった「自由な思考」と「理にかなった思考」の両立が可能になる．その具体的な特長として，以下が挙げられる．

1．整理しやすい
　　・デザイン要素間の関係が明らかにできる
　　・アイデアの違いを明らかにできる
　　・思考の過程を明らかにできる
2．使いやすい
　　・さまざまな対象領域でも使える
　　・自由なやり方で使える
　　・他者とのコラボレーションに使える
3．発想しやすい
　　・新たな価値を生むアイデアを発想できる
　　・場に適し，場を創るアイデアを発想できる
　　・シーズを生かしたアイデアを発想できる

M Method —For Coexistence of "Unrestricted Thinking" and "Rational Thinking"

The "**M method (multispace design method)**" is proposed as a design method based on "multispace design model" (figure 15). This "M method" is a usable thinking method that can be used in various fields including design, business, planning department, research, and in everyday life (figure 16). This trait comes from the point where the user thinks by "being able to adapt to individual approach and facilitate modeling and idea generation through analysis using the perspective of multispace". This enables the user to conduct both "unrestricted thinking" and "rational thinking" in innovating products with large scale structures and systems. The specific characteristics of M method are given below.

1. Organization
    - enables the "clarification of the relationships between design elements"
    - enables the "distinction between differences in ideas"
    - enables the "clarification of the thinking processes"
2. Usability
    - can be applied to "diverse domains"
    - can be applied to "individual design approaches"
    - can be applied to "multi-person collaborations"
3. Ideation
    - can generate ideas "to create new values"
    - can generate ideas "appropriate to and to create new circumstances"
    - can generate ideas "using innovation seeds"

「Mメソッド」は，すでにメガネやUSBメモリのような小物から，切削加工システム，半導体生産システム，自動車やそのサービスシステムを含む大規模な交通システムまで，さまざまなデザイン対象に利用されており，その有効性が立証されている．

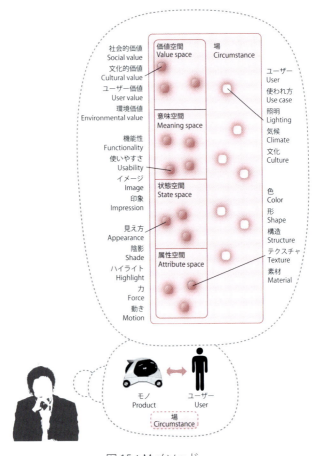

図15：Mメソッド
Figure 15 : M method

This "M method" has already been applied in innovating various effective products from glasses and USB to large scale transportation system, semiconductor production system, and vehicle service system.

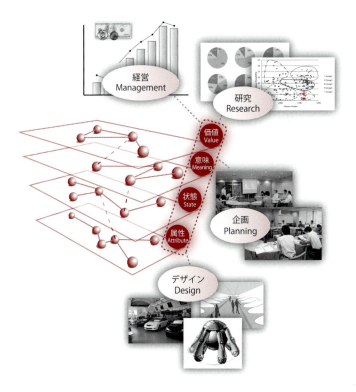

図 16：M メソッドの様々な対象領域
Figure 16 : Diverse application domains of M method

ものづくりへの期待
──新たな価値創造の実現のために

　今後も，ものづくりは，大規模化・複雑化が進むことは必至である．そのため，「自由な思考」と「理にかなった思考」の両立が不可欠であり，それによるイノベーティブでかつ完成度の高い新たな価値の創造が強く求められている．

　今，デザイン科学という新たな学問が急成長している．そして，その所産である多空間のデザイン思考：Mメソッドを活用することで，独自の方法論を構築し，世界を先導する新たな価値創造の実現が期待されている．

Expectation Toward Japan's Manufacturing —For Realization of New Value Creation

From here on, the promotion of a larger scale and more complex manufacturing is inevitable. Therefore, the combination of "rational thinking" and the "unrestricted thinking" is important to create a more innovative and perfected new value.

Currently, a new study called the "design science" is rapidly growing. It is believed that by employing the design thinking of "multispace" (M method), which is based on design science, enables the innovation of a unique methodology and the realization of a new value creation which can lead the world.

第10の知恵

「マルチコンカレント」でリードする

第4次産業革命がもたらす
「早い者勝ち時代」に何をなすべきか？

10th Guideline

Leading by "Multi-Concurrent"

What Should We Do in this Period of "First-Come, First-Serve" Brought by the 4th Industrial Revolution?

第 4 次産業革命

第 4 次産業革命に向けたさまざまな活動がなされているのは，ご存じのとおりである．これまで，18 世紀後半の第 1 次産業革命では石炭・蒸気機関の活用による機械化が行われ，20 世紀初頭の第 2 次産業革命では電気エネルギーの利用による大量生産が可能になった．続いて，20 世紀後半の第 3 次産業革命ではコンピュータによる自動化とそれに伴う大量生産技術が進化した．

そして第 4 次産業革命では，AI（人工知能），IoT（Internet of Things），ビッグデータなどの活用により，ロボットや工作機械をはじめ，あらゆるものがネットワーク化されることで，産業が新たなフェーズに進むとされている．その具体的な在り方にはさまざまな議論があるが，いずれにしても，今後，産業がそのような方向に進むことは間違いないであろう（図 17 参照）．

その結果，工場生産が自動化されるのみならず，工場と店舗や消費者が直接つながることで，大量生産のみならずカスタマイズ生産の技術革新が期待されている．また，それを発展させることで，新ビジネスモデルの構築など，さまざまな可能性が検討されている．たとえば，モノづくり産業のみならずサービスなどのモノづかい産業とのシナジーを発揮する「**モノづくり×モノづかい産業**」のためのビジネスモデルである．今，それらに向けた戦略的技術開発は，重要かつ急務であるといえる．

The 4th Industrial Revolution

As you may know, various activities toward the **4th Industrial Revolution** have been going on. In the 1st Industrial Revolution of the late 18th century, the mechanization with the adoption of coal and steam engine occurred. Next, in the 2nd Industrial Revolution of the early 20th century, the mass production with the usage of electricity was achieved. Then, in the 3rd Industrial Revolution of the late 20th century, the automation and mass production technologies with the adoption of computers successfully evolved.

In the 4th Industrial Revolution, it is believed that the manufacturing of the world will move to a next stage, due to the networking of various things (ex. robots and machines) with the utilization of AI, IoT, and big data. There may be some arguments on the way of how the future manufacturing should be. However, it is without a doubt, that industries will move toward the 4th Industrial Revolution (figure 17).

As a result, the automation of factory production directly connects the factories, shops, and customers, which leads to the expectation on innovation of the mass production and also the customization production. In addition, by developing this new area of production, various possibilities including the construction of a new business model is being discussed. For example, the "**Product creation × Product usage industry**", a business model which shows synergy between the product creation industry and product usage industry (such as services). Currently, strategic technology development is under an important and urgent situation.

18 世紀後半
Late 18th century

第 1 次産業革命
1st Industrial Revolution

石炭・蒸気機関の活用による機械化

Machanization with adoption of steam engines and coals

20 世紀初頭
Earty 20th century

第 2 次産業革命
2nd Industrial Revolution

電気エネルギーの利用による大量生産

Mass production with usage of enectricity

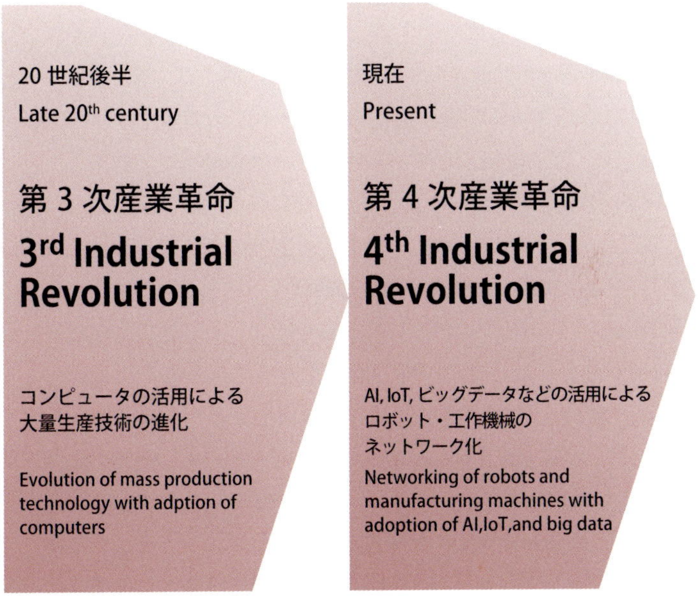

20 世紀後半
Late 20th century

第 3 次産業革命
3rd Industrial Revolution

コンピュータの活用による
大量生産技術の進化

Evolution of mass production technology with adption of computers

現在
Present

第 4 次産業革命
4th Industrial Revolution

AI, IoT, ビッグデータなどの活用による
ロボット・工作機械の
ネットワーク化

Networking of robots and manufacturing machines with adoption of AI, IoT, and big data

図 17：産業革命の推移

Figure 17 : Change of the Industrial Revolution

「アイデア勝負の時代」から
「早い者勝ちの時代」へ

　このように，第4次産業革命がもたらすものにはさまざまな可能性が考えられるが，一つ気になる点がある．それは，「**早い者勝ち時代**」への突入である．なんだか，イヤな響きである．効率化の極みを要求されるストレス社会のにおいがする．しかしながら，現在のAIやIoTの進化を見れば，すべての領域とはいわないが，モノづくりがその方向に向かうことは否めないであろう．

　これまでのモノづくりはというと，高い技術力を背景にしつつも，新しいアイデアで勝負してきた感がある．長きにわたり，人間の発想力による「**アイデア勝負の時代**」であったといえる．しかし，そのアイデアが，将来的にはAIにより，瞬時に生まれる可能性が高い．なぜなら，AIが発想力を持ちはじめているからである．近年，AIが将棋や囲碁でプロ棋士に勝ち，料理のレシピを考え，小説を書きはじめていることなどからも，容易に想像できるのではないだろうか．しかも，情報収集能力が極めて高いIoTが，瞬時に世界中の情報を集めてくれて，その発想を加速させる．そのため，AIが人間よりもアイデア発想力に勝る時代がいつかはやって来ると言わざるをえないのではないだろうか．

　そのことを考えると，第4次産業革命の時代では，いかに新たなアイデアを生み出すかよりも，AIやIoTなどにより瞬時に生み出されたアイデアをいかに早く具現化し，生産，市場に投入するかがカギを握るということになる．第4次産業革命は，「アイデア勝負の時代」から「早い者勝ちの時代」への移行をもたらす可能性が高いといえる．これは，これまでAI研究に関わってきた私自身の実感でもある．

From Period of "Ideas are Everything" to Period of "First-Come, First-Serve"

As you can see, there are many kinds of possibilities that the 4th Industrial Revolution possesses. However, there is one point that I am concerned about, which is the shift to the "**period of first-come, first-serve**". This phrase seems to gives an unpleasant feeling. I smell a stressful society, demanded to perform extreme efficiency, from this phrase. However, by observing the evolution of current AI and IoT, it is definite that product creation in most fields will see the period of "first-come, first-serve".

With that in mind, in the 4th Industrial Revolution, rather than generating new ideas, but the speed in realizing the idea by using the AI or IoT and bring it into the market holds the key in success. Therefore, you can say that there is a high possibility that the period will shift from the "**period of ideas are everything**" to the period of "first-come, first-serve" in the 4th Industrial Revolution. In fact, as one of the researcher on AI, this is my actual feeling now.

マルチコンカレントの必要性

　では，そのような「早い者勝ちの時代」に向けて，我々は今，何をなすべきか？これにはさまざまな課題と方策が考えられるが，ここでは，従来，開発期間短縮に有効であった**コンカレントエンジニアリング（CE）**に注目する.

　コンカレントエンジニアリングは，開発の上流過程と下流過程を同時進行させる手法である．1970年代前半に米国が，日本の慣行であった根回しによる擦り合わせ型開発を参考にし，製品開発システムとして構築したことで知られている．1975年に発表されたフォードのトーラス車がそれを用いて開発され，注目を浴びた．市場に投入されたトーラス車は非常に品質も良く，高い評価を得た．そして，この車が，実はコンカレントエンジニアリングによるものであったことに気がついた日本の自動車メーカは，慌てて，自らの慣行を参考にシステム化されたコンカレントエンジニアリングを学び，導入を急いだ．当時，自動車メーカで開発の真っただ中にいた私は，今でもその時の様子をはっきり覚えている．

　さて，先に示したように，第4次産業革命がもたらす「早い者勝ちの時代」においては，このコンカレントエンジニアリングをさらに発展させることが有効であると考える．単に上流過程と下流過程の同時進行にとどめるのではなく，さまざまな同時進行に拡張する**マルチコンカレントエンジニアリング（MCE）**の構築が必要であろう（図18参照）．たとえば，ハードウェアとソフトウェア，トータルシステムとサブシステム，さらには，モノづくり（製品開発システム）とモノづかい（サービスシス

Importance of Multi-Concurrent Engineering

Then, for such period of "first-come, first-serve", what is it that we can do? There are many problems and solution to this question. Though, in this section, we will be paying attention to **concurrent engineering (CE)**, which is said to be effective in shortening development period.

The concurrent engineering is a method to let you progress with upstream development and downstream development at the same time. In 1970, the United States developed this product development system by referring to "nemawashi", which is the way of gaining consensus in Japan. Then, this system attracted attention in 1975, when Ford Taurus was announced using it for development. The Taurus that was put into the market was high quality and had high reputation. As soon as the Japanese automobile company figured out that the Taurus was developed using the concurrent engineering, they rushed to study to apply the systematized concurrent engineering developed using their general manner in Japan. Indeed, I experienced that moment while working in the automobile industry, and to this date I remember all the details.

In fact, it is thought effective to develop and apply this concurrent engineering in the period of "first-come, first-serve" (the 4th Industrial Revolution). The construction of the **multi-concurrent engineering (MCE)** that can lead to expansion of the field toward various direction is necessary, and not just to settle with maintaining the upstream and downstream progresses concurrently (figure 18). As an example, hardware/ software,

テム）など，さまざまなコンカレント（同時進行）である．さらに，マルチコンカレントエンジニアリングの対象は，開発する製品の要素にとどまらない．作り手側や使い手側を含むさまざまなコンカレント，たとえば，川上企業と川下企業間，開発者と消費者を含むすべてのステークホルダー間はもちろんのこと，人間の知能と人工知能とのあいだのコンカレントなどの在り方も問われる．そして，それらを AI や IoT などのコンピュータを介して，同時に開発，製造，販売，サービスを行う．それにより，開発・製造からサービスまでの超期間短縮化を図り，「早い者勝ち時代」に対応しようとするものである．

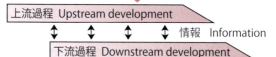

total system/ subsystem, and product creation (product development system)/ product usage (service system) are all concurrent (progresses concurrently). Furthermore, the object of multi-concurrent engineering does not remain only in product development factors. The multi-concurrent includes the maker/ user; for example, upper stream company/lower stream company, developer/ consumer, and human intelligence/ artificial intelligence. However, the ways of how these concurrent should be is questioned now. In addition, the multi-concurrent will utilize AI and IoT to concurrently develop, manufacture, sale, and provide services. These lead to the shortening of time from developing to manufacturing to providing services, and will correspond to the period of "first-come, first-serve".

〈 マルチコンカレントエンジニアリング (MCE) 〉
Multi-concurrent engineering

以下を同時進行させる
Progress the factors below concurrently

- トータルシステム／サブシステム
  Total system / Sub system
- ハードウェア／ソフトウェア
  Hardware / Software
- モノづくり（製品開発システム）／モノづかい（サービスシステム）
  Product creation (product development system) /
  Product usage (service system)
- 川上企業／川下企業
  Upper stream company / Lower stream company
- 開発者／消費者
  Developer / Consumer
- 人間の知能／人工知能
  Human intelligence / Artificial intelligence

図18：マルチコンカレントエンジニアリング（MCE）
Figure18 : Multi-concurrent engineering

従来の風土や価値観を見直すチャンス

　マルチコンカレントエンジニアリングの効果は，「早い者勝ち時代」への対応としての開発期間の短縮，開発費の削減，市場ニーズへの迅速な対応などだけではない．企業におけるこれまでの製品開発システム，さらには，企業の風土や価値観が時代に合致しているかを再考する良い機会にもなる．

　マルチコンカレントエンジニアリングは，製品の要素，作り手側や使い手側の要素などあらゆる要素を同時進行させる．そのことは，言い換えれば，あらゆる要素が時間軸で平等に扱われるということになる．しかし，企業には，継承されてきた製品に関わる優先要素とそれに伴う開発システムが現存する．そして，それらの優先要素や製品開発システムは，その製品開発に対する思想，さらには企業固有の風土や価値観などが反映されたものである．

　しかし，マルチコンカレントエンジニアリングの実行のためには，そのような優先要素をすべていったんリセットさせる必要がある．そのため，これを導入するためには，既存の製品開発システムや思想，風土や価値観までの再考が問われる．このことは，企業においては大変な作業となるが，これまで継承してきたそれらが，現在あるいは将来の時代に適合しているかを見つめ直す良いチャンスとなる．

　一般に，企業における製品開発システムやそれを生み出す風土・価値観は，それ自体，急激に変化しにくい．時代の要請に合わせ，時間軸でイナーシャ（慣性力）を持ちながら徐々に変化する．そのため，急激な時代変化があっても，それに合致さ

Chance to Look Back on Conventional Culture and Sense of Value

The effect of multi-concurrent engineering corresponds to the period of "first-come, first-serve", leading to shortening of development span, reducing development costs, and coping speedily to the market needs. In addition, there are more effect of multi-concurrent. Multi-concurrent Engineering brings good opportunity to rethink about the company's current product development and to understand if the company's cultures and sense of values fits with the present age.

The multi-concurrent engineering enables concurrent progress of various factors such as the product's factor, the maker and the user's factors. In other words, all kinds of elements can be evenly treated in the same timeaxis. However, the current companies still used passed on development system that consists product development system with priority elements. In addition, such priority elements and product development systems reflects the company's thoughts toward product development, cultures, and sense of values.

Still, in order to perform multi-concurrent engineering, the company must once reset all priority elements. Therefore, in order to apply the multi-concurrent engineering, the company must also rethink about their current development system, thoughts, culture and sense of values. Although, it may sound difficult, this opportunity is a chance to reconsider if the system developed up until now fits in the current or the future era.

In general, it is extremely difficult to change the company's current product development systems , cultures, and sense of

せた大きな変革の必要性に気づきにくい．それは，まるで「ゆでガエル」のごとくである．また，気づいたとしてもその実行には難しい側面がある．そのため，今日のような急激な変化の時代には，危険な状態にあるかもしれない．マルチコンカレントエンジニアリングの導入は，そのような改革の必要性を改めて問う「機」となり，その意味でも，今，必要とされているのかもしれない．

values. The change occur gradually, with the inertial force in the timeaxis, and adapting to the requirement of the era. Therefore, even if there were a drastic era of change, we may not notice the necessities of the revolution that corresponds to the change, like a "boil frog". Thus, under the current drastic change of era, it may be danger to achieve. The application of multi-concurrent engineering is necessary in a sense because, it could lead to gaining a "chance" to rethink the importance of the revolution we are about to confront.

付録

1. デザインサイエンス
2. 多空間デザインモデル（M モデル）
3. AGE 思考モデル

APPENDIX

1. Design Science

2. Multispace Design Model: M Model

3. AGE Thinking Model

1. デザインサイエンス

デザインサイエンスは，デザインという人の創造的行為における法則性の解明と，デザイン行為に用いられる様々な知識の体系化を狙いとする学問である．このため，デザインサイエンスは，プロダクトデザイン，建築デザイン，情報デザイン，サービスデザインなどの様々なデザインの分野において共通の基盤となる．以下に，デザインサイエンスの概念や位置づけの変遷を述べる．

デザインサイエンスという用語を初めて用いたのは，1960年代に著書の中で"宇宙船地球号"という概念を示した建築家・思想家の B. Fuller であり，その後現在までに，多くの研究者によってデザインサイエンスに関する議論が行われてきた．1970年代に，F. Hansen は，デザインサイエンスの目標を"デザイン行為における法則の認識と規則の構築"と位置づけた．1980年代には，V. Hubka と W. E. Eder が，デザインサイエンスを Hansen よりも広い概念で捉え，"デザイン領域における知識の集合やデザイン方法論の概念なども含むもの"と位置づけた．

1990年代に，N. Cross は，デザインサイエンスを"デザイン対象に対して組織化・合理化されたシステマティックなアプローチ"と表現し，科学的知識を活用するにとどまらない科学的行為としてデザインを捉えた．なお，Cross はデザインサイエンスとデザイン学の相違についても言及し，デザイン学を"科学的な探求手法を通じてデザインに関する我々の理解を改善しようとする一連の研究"と位置づけた．

2000年代に，Y. Matsuoka は自らの主宰するデザイン塾にお

1. Design science

Design science is an academic discipline focused on the clarification of the laws that govern design as a human act of creativity as well as the systemization of the knowledge used in designing. Design science therefore functions as the foundation for all design fields: including product design, architectural design, information design, service design and so on. The followings present a brief overview of the principle concepts of design science and describe changes in the way that design science has been defined over time.

The term "design science" was first used at the beginning of the 1960s, in a work by Buckminster Fuller, an architect and thinker who developed the idea of "Spaceship Earth". Since then, design science has been widely debated by many researchers; it still is today. In the 1970s, Friedrich Hansen defined the aim of design science as being to "recognize laws of design and its activities, and develop rules". In the 1980s, Vladimir Hubka and Wolfgang Ernst Eder proposed that design science in fact had a wider scope than Hansen's definition allowed, arguing rather that it "comprises a collection (a system) of logically connected knowledge in the area of design, and contains concepts of technical information and of design methodology".

In the 1990s, Nigel Cross suggested a new definition for design science, describing it as implying "an explicitly organised, rational and wholly systematic approach to design"; his definition proposed design science to be a scientific activity itself, one which utilizes more than just scientific knowledge. Cross

いて，デザインサイエンスを"デザイン行為における法則性の解明およびデザイン行為に用いられる知識の体系化を目指す学問"と表現し，デザインに関わるあらゆる事象の科学的な解明を目指すデザイン学における中核を成すものと位置づけた．

デザインサイエンスの枠組みは，**デザイン知識**と**デザイン行為**の2つで構成されると考えられている．ここで，デザイン知識は，科学的知識のような**客観的知識**と個人的な経験知のような**主観的知識**から成る．一方で，デザイン知識に基づいて行われる**デザイン行為**は，**デザイン実務，デザイン方法，デザイン方法論，デザイン理論**の4つの階層から成る（図A）．デザイン行為の4階層においては，上位の階層になるほど特殊性・具体性が増していき対象に依存する特徴がある．反対に，下位の階層になるほど一般性・抽象性が増していき対象に依存しない特徴がある．

次に説明する**多空間デザインモデル**は，この枠組みの最下層に位置するデザイン理論に関するモデルである．

also discussed the difference between design science and the science of design, defining the science of design as "that body of work which attempts to improve our understanding of design through 'scientific' (i.e., systematic, reliable) methods of investigation".

In the 2000s, Yoshiyuki Matsuoka defined design science as an "academic discipline focused on the clarification of the laws that govern design as a human act of creativity as well as the systemization of the knowledge used in designing". This definition situates design science at the core of the science of design, which seeks to clarify our understanding of all the matters related to design.

The **framework of design science** is made up of **design knowledge** and **designing**. Design knowledge is made up of **objective knowledge**, as exemplified by scientific knowledge, and of subjective knowledge, such as individual experience. However, designing as an act conducted on the basis of design knowledge, is made up of the following four levels: **design work**, **design method**, **design methodology**, and **design theory** (Figure A). The higher the level, the greater the concreteness and particularity, and the more designing will closely depend upon the design objective. Conversely, the lower the level, the greater the universality and abstractness; this means a lesser dependence on the design objective.

The following section will introduce the **multispace design model**. This model is focused on design theory, which occupies the lowest level of the design framework.

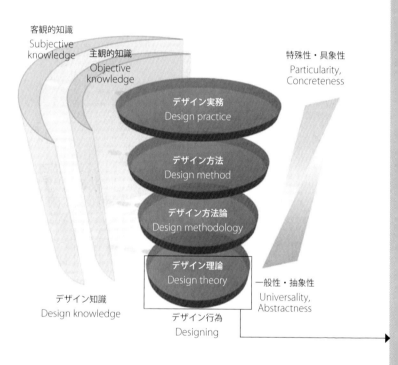

図A：デザイン科学の枠組み

Figure A : Framework of design science

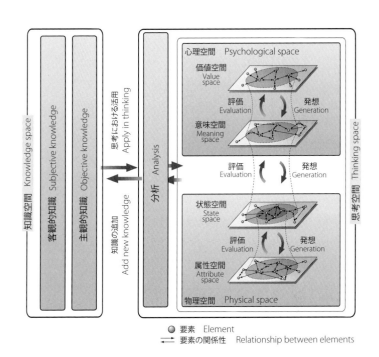

図B：多空間デザインモデル（M モデル）

Figure B : Multispace Design Model (M model)

APPENDIX

2. 多空間デザインモデル（M モデル）

 多空間デザインモデルとは，複数の空間にデザイン行為とそれに用いるデザイン知識を記述することで，デザイン行為をモデル化したものである．ここで，空間とは，デザインの思考において用いられるデザイン要素やデザイン知識の集合であり，この空間を用いてそれらの関係性を記述することができる．多空間デザインモデルでは，デザイン行為が**思考空間**に，デザイン知識が**知識空間**にそれぞれ記述される．

思考空間：デザイン行為が記述された空間．
知識空間：デザイン行為に用いる知識が記述された空間．

 思考空間では，多空間とデザイン思考によりデザイン行為が記述される．以下に，多空間について説明する．なお，デザイン思考については後述の「**AGE 思考モデル**」で説明する．

 多空間における各空間には，デザインを行う際に用いる様々な言葉や画像などで表現されるデザイン要素とそれらの関係性が記述される．多空間は，心理的なデザイン要素とそれらの関係性が記述された**心理空間**と，物理的なデザイン要素とそれらの関係性が記述された**物理空間**に分けられる．さらに，心理空間は**価値空間**と**意味空間**に，物理空間は**状態空間**と**属性空間**に分けられる．

2. Multispace Design Model (M model)

The **multispace design model** enables the modeling of designing and the design knowledge used therein onto multiple spaces. "Space" here is used to refer to a collection of design elements and design knowledge used in design thinking. The model allows the designer to express the relationships that exist between these spaces. In the multispace design model, designing is expressed in terms of **thinking space** and design knowledge in terms of **knowledge space**.

Thinking space: space describing designing.
Knowledge space: space describing the knowledge used in designing.

Thinking space is the space in which the designing resulting from the multispace and design thinking approach is expressed. The following section introduces the concept of the multispace. An introduction to design thinking follows, in the section entitled "**AGE thinking model**".

Each space within the multispace is used to express the design elements, expressed by various words and images, used in the course of design and to express the relationships between those elements. The multispace can be divided into **psychological space**, in which psychological design elements and their inter-relationships are mapped out, and **physical space**, which comprises physical design elements and their inter-relationships. These spaces can be further subdivided: psychological space into **value space** and **meaning space**; physical space into **state space** and **attribute space**.

心理空間：心理的要素とその要素間の関係性が記述された空間.
物理空間：物理的要素とその要素間の関係性が記述された空間.

価値空間：社会的価値，文化的価値，個人的価値など多様な価値を表すデザイン要素が記述された空間.
意味空間：価値を実現するための機能性やイメージなどのデザイン要素が記述された空間.
状態空間：意味を実現するための動きや陰影などの要素が記述された空間．なお，これらの要素はヒトや環境などの場に依存する物理的特性である.
属性空間：状態を実現するための人工物の形，色，素材などの要素が記述された空間.

　知識空間では，デザインを行う際に用いる様々な知識が記述される．知識は，自然科学，人文科学，社会科学などに基づく一般性のある客観的知識と，デザイナーの個人的な経験や地域性などに基づく主観的知識に分けられる．一般に，知識には言語，図表，数式などの記号により表現可能な**形式知**と，言語などの記号では表現が難しい**暗黙知**があるが，一般性を有する形式知は**客観的知識**に，一般性を有さない形式知または暗黙知は**主観的知識**に記述される．以上から，多空間デザインモデルによりデザイン行為は，図B（P151）のように記述される.

Psychological space: space consisting of psychological elements and the relationships between them.

Physical space: space consisting of physical elements and the relationships between them.

Value space: space consisting of design elements expressing diverse values, including social value, cultural value, and individual value.

Meaning space: space consisting of design elements required to express values, such as functionality and image.

State space: space consisting of design elements that are required to manifest meaning, such as movement or shadow and are physically characterized by a reliance on circumstance, such as humans or environments.

Attribute space: space consisting of artificial elements such as shape, color, and material required to bring about state.

Knowledge space is where the various types of knowledge used in the design process are described. Knowledge here can be divided into objective knowledge, based on fields such as the natural sciences, the humanities, and social sciences, and subjective knowledge, which is particular to the experiences of the individual designer and his or her regional location. In general, knowledge includes **explicit knowledge**, codified through symbols, such as language, graphs, and math ematical formula, and **tacit knowledge**, which is more difficult to express symbolically, through language for example. That explicit knowledge that is general is expressed in **objective knowledge**; individual knowledge and tacit knowledge are expressed in **subjective knowledge**.

The designing undertaken through the multispace design model can be expressed as shown in Figure B (P151).

3. AGE 思考モデル

デザイン行為は多空間デザインモデルにおいて **AGE 思考モデル**により記述することができる．AGE 思考モデルは**分析（帰納）**，**発想（仮説形成）**，**評価（演繹）**の 3 つで構成される．

分析（帰納）：デザイン問題に関連する様々な要素間の関係性を明確にすることで，その背景に存在する一般則を導く行為（特殊性を有する個別の事象から一般性を有する法則を導く推論行為）．
発想（仮説形成）：与えられたデザイン問題を解決するデザイン案を導く行為（個別の事象を最も適切に説明しうる結論を導く推論行為）．
評価（演繹）：一般則に基づきデザイン問題に関連するさまざまな要素の位置づけを明確にする行為（一般性を有する前提から特殊性を有する結論を導く推論行為）．

図 C は，多空間デザインモデルにおける状態空間と属性空間を例として，AGE 思考モデルとの対応を記述したものである．まず，デザイナーや設計者はデザイン対象を分析する．つぎに，分析しながら与えられたデザイン問題を解決するようなデザイン案を発想する．そして，発想したデザイン案を評価する．その結果，デザイン案が与えられたデザイン問題を解決できると判断された場合は最終的なデザイン案となり，解決できないと判断された場合は再び分析し，新たなデザイン案を発想する．

このように，デザイン行為は，分析，発想，評価という 3 つのデザイン思考を繰り返すことにより進められる．

3. AGE thinking model

It is possible to model designing through **AGE thinking model** using the multispace design model. AGE thinking model can be composed of three types: **design problem analysis (induction)**, **idea generation (abduction)**, and **idea evaluation (deduction)**.

Design problem analysis (induction): analysis clarifies the relationships among the various elements related to the design problem and leads to the establishment of general rules providing the context to the design problem (reasoning activity leading to general rules with universality from individual cases characterized by particularity).

Idea generation (abduction): generation leads to design ideas as potential solutions to the design problem presented (reasoning activity leading to conclusions with the potential to be the most convincing explanation for individual cases).

Idea evaluation (deduction): evaluation clarifies the positioning of the various elements related to the design problem according to general rules (reasoning activity leading toconclusions with particularity from conditions with universality).

Figure C takes the example of the state space and attribute state of the multispace design model, and indicates how these states correspond to AGE thinking model. First, the designers working on the design problem must analyze the design objective. As they analyze they must generate design ideas representing potential solutions to the design problem at hand. Next, they

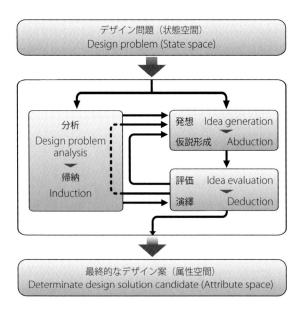

図C：AGE 思考モデル

Figure C : AGE thinking model

must evaluate the design ideas that have been generated. As a result, those design ideas determined to solve the design problem at hand will be selected as final design solution, while design idea rejected are analyzed again, resulting in the generation of new design ideas.

In this way, designing is progressed through the iteration of three types of thinking: design problem analysis, idea generation, and idea evaluation.

参考文献

- 松岡由幸, 他:『製品開発のための統計解析学』, 共立出版, 2006

- 松岡由幸:『デザインサイエンス ── 未来創造の"六つ"の視点』, 丸善, 2008

- 松岡由幸, 他:『もうひとつのデザイン ── その方法論を瀬見恵美に学ぶ』, 共立出版, 2008

- 松岡由幸, 宮田悟志:『最適デザインの概念』, 共立出版, 2008

- 松岡由幸:『図解 形状設計ノウハウハンドブック ── デザイン科学が読み解く熟練設計者の知恵と工夫』, 日刊工業新聞社, 2010

- 松岡由幸:『タイムアクシス・デザインの時代 ── 世界一やさしい国のモノ・コトづくり』, 丸善出版, 2012

- 松岡由幸, 他:『M メソッド ── 多空間のデザイン思考』, 近代科学社, 2013

- 松岡由幸, 他:『創発デザインの概念』, 共立出版, 2013

- 松岡由幸, 加藤健郎:『ロバストデザイン ──「確かさ」に対して頑強な人工物の設計法』, 森北出版, 2013

- 青木弘行, 松岡由幸, 他:『プラスチックの逆襲』, 丸善プラネット, 2017

REFERENCES

- Yoshiyuki MATSUOKA, *et. al.*: *Statistical Analysis for Product Development*, Kyoritsu Shuppan, 2006
- Yoshiyuki MATSUOKA, *et. al.*: *Another Type of Design*, Kyoritsu Shuppan, 2008
- Yoshiyuki MATSUOKA, Satoshi MIYATA: *Concept of Optimum Design*, Kyoritsu Shuppan, 2008
- Yoshiyuki MATSUOKA: *Design Science – "Six Viewpoints" for the Creation of Future*, Maruzen, 2010
- Yoshiyuki MATSUOKA: *Handbook of Shape Design Knowhow*, Nikkan Kogyo Shinbun, 2010
- Yoshiyuki MATSUOKA: *Dawn of Timeaxis Design*, Maruzen, 2012
- Yoshiyuki MATSUOKA, *et. al.*: *M Method –Design Thinking on Multispace*, Kindai-Kagaku-Sha, 2010
- Yoshiyuki MATSUOKA, *et. al.*: *Concept of Emergent Design*, Kyoritsu Shuppan, 2013
- Yoshiyuki MATSUOKA, Takeo KATO: *Robust Design*, Morikita Shuppan, 2013
- Hiroyuki AOKI, Yoshiyuki MATSUOKA, *et. al.*: *The Avenging Plastics*, Maruzen Planet, 2017

人名索引

Cross, N. 146
Eder, W. E. 146
Fuller, B. 146
Hansen, F. 146
Hubka, V. 146
Kurzweil ,R. 62
Matsuoka, Y. 146

索　引

数字・欧文
　２次産業　22
　３次産業　22
　３Ｄプリンター　24
　６次産業　22, 38
　AI　58
　AGE 思考モデル　64, 152, 156
　CE　136
　IoT　68
　MCE　136
　Mメソッド　122
　Mモデル　118
　UXデザイン　12

あ アイデア勝負の時代　134
　アイデア発想　68
　アナロジー（類推）型発想　68
　暗黙知　154
い 意味　118
　意味空間　152
え エキスパートシステム　58
　演繹　64, 156
お 驚き　10
　オブジェクト型　4
　オントロジー　70
か 学習型　36
　カスタマイズ型　34

　仮説形成　64, 156
　可塑モデル　48
　価値　118
　価値空間　152
　価値実感期　50
　価値成長期　50
　価値成長デザイン　28, 48
　価値成長メカニズム　50
　価値定着期　50
　価値伝承　50
　価値発見期　50
　感動　2
　感動のメカニズム　10
き 帰納　64, 156
　客観的知識　148, 154
　共感　10, 12
　共創システム　72
　極めるマインド　54, 90
け 形式知　154
こ コトづくり　38
　コンカレントエンジニアリング　136
　コンテクスト型　4
さ サービス科学　50
　サービス技術　36, 50
　サービス工学　50
　最適デザイン　80

INDEX OF PERSONS

Cross, Nigel 147
Eder, Wolfgang Ernst 147
Fuller, Buckminster 147
Hansen, Friedrich 147

Hubka, Vladimir 147
Kurzweil, Ray 63
Matsuoka, Yoshiyuki 149

INDEX

3D printer 25
4th Industrial Revolution 131
6th industry 23, 39

A abduction 65, 157
accustoming type 33
AGE thinking model 65, 153, 157
AI 59
amazement 11
analogical generation type 69
analysis 65
Artificial Intelligence 59
attribute 97, 119
attribute space 153

B background type 5
bio-inspired technology 37, 51
bottom up 77
bottom-up design 81

C CE 137
character extraction 59
circumstances 19
co-creation system 73
concurrent engineering 137
context type 5
customizing type 35

D deduction 65, 157

deep emotion 3
deep learning 59
design for "creation" 19
design for "usage" 19
designing 149
DesignJuke 119
design knowledge 119, 149
design method 149
design methodology 149
design problem analysis 157
design science 119
design theory 149
design thinking 119
design work 149

E emergence 77
emergent design 81
empathy 11
evaluation 65
expert system 59
explicit knowledge 155

F framework of design science 149

G generation 65

I idea evaluation 157
idea generation 69, 157
induction 65, 157
integrative generation type 69

し	思考空間 152		デザイン方法 148
	自然変化型 30		デザイン方法論 148
	主観的知識 148, 154		デザイン理論 148
	状態 96, 118	と	統合型発想 68
	状態空間 152		特徴抽出 58
	シンギュラリティ 62		トップダウン型のデザイン 80
	人工知能 58		トレードオフ問題 106
	心理空間 152	な	馴染み型 32
そ	想定外 104	に	ニューラルネットワーク 58
	想定しなかった想定外 104	の	ノウハウの伝承 100
	想定できなかった想定外 104	は	場 18
	創発 76		バイオ・インスパイヤード技術 36, 50
	創発デザイン 80		創めるマインド 76, 90
	属性 96, 118		バックグラウンド型 4
	属性空間 152		発想 64, 156
	「育つ」技術 50		早い者勝ち時代 134
	「育てる」技術 50	ひ	非定常モデル 48
た	タイムアクシス 44		評価 64, 156
	タイムアクシスストラテジー 72	ふ	分析 64, 156
	タイムアクシスデザイン 12, 22, 48		物理空間 152
	第4次産業革命 130	ほ	ボトムアップ 76
	多空間デザイン法 122		ボトムアップ型のデザイン 80
	多空間デザインモデル 118, 148, 152	ま	マルチコンカレントエンジニアリング 136
	多層構造 58		マルチタイムスケールモデル 48
ち	知識空間 152	め	メタファー 70
つ	「つかう」デザイン 18	も	モノづかい 20, 22
	「つくる」デザイン 18		モノづくり 16, 20, 22, 38
て	ディープラーニング 58		モノづくり×モノづかい産業 130
	デザイン行為 148	ゆ	ユーザーエクスペリエンスデザイン 12
	デザインサイエンス 118	ろ	ロバスト 108
	デザインサイエンスの枠組み 148		ロバストデザイン 106, 108
	デザイン思考 118		
	デザイン実務 148		
	デザイン塾 118		
	デザイン知識 118, 148		

K
- IoT 69
- knowhow tradition 101
- knowledge space 153

L
- learning type 37

M
- MCE 137
- meaning 119
- meaning space 153
- mechanism of deep emotion 11
- metaphor 71
- mind to originate 77, 91
- mind to pursue 55, 91
- M method 123
- M model 119
- multi-concurrent engineering 137
- multilayered 59
- multispace design method 123
- multispace design model 119, 149, 153
- multi-timescale model 49

N
- natural changing type 31
- neural network 59
- non-steady model 49

O
- object type 5
- objective knowledge 149, 155
- ontology 71
- optimum design 81

P
- period of first-come, first-serve 135
- period of ideas are everything 135
- physical space 153
- plasticity model 49
- product creation 17, 21, 23, 39
- Product creation × Product usage industry 131
- product usage 21, 23
- psychological space 153

R
- robust design 107, 109
- robustness 109

S
- secondary sector of industry 23
- service engineering 53
- service science 53
- service technology 37, 51
- Singularity 63
- state 97, 119
- state space 153
- subjective knowledge 155

T
- tacit knowledge 155
- TaD 13
- technology "to grow" 51
- technology "to nurture" 51
- tertiary sector of industry 23
- thinking space 153
- timeaxis 45
- timeaxis design 13, 23, 49
- timeaxis strategy 73
- top-down design 81
- trade-off problem 107

U
- unexpected 105
- user experience design 13
- UX design 13

V
- value 119
- value creation 39
- value discovery phase 51
- value establishment phase 51
- value growth design 29, 49
- value growth mechanism 51
- value growth phase 51
- value realization phase 51
- value space 153
- value tradition 51

W
- was not assumed 105
- was unable to be assumed 105

著者紹介

松岡由幸

慶應義塾大学教授,デザイン塾主宰
山口県下関生まれ.
専門は,デザイン科学,設計工学,製品開発システム論.
多空間デザインモデル(Mモデル),Mメソッド,AGE思考モデル,創発デザインなどのデザイン科学の基礎を構築.また,デザイン・設計に時間軸を組み込む「タイムアクシスデザイン」も提唱.これらを用いて,多くの企業が新製品やシステムの開発を実施.
日本デザイン学会(会長),日本設計工学会(副会長),日本工学会(フェロー),日本機械学会(フェロー),基礎デザイン学会(監事),CG ATRS協会(委員),日本インダストリアルデザイナー協会,ASME,IEEE,Design Society.
イリノイ工科大学デザイン研究所客員フェロー,経済産業省・中小企業庁などの各種委員,横断型基幹科学技術研究団体連合理事,機械工業デザイン賞専門審査委員を歴任.

AUTHOR NOTE

Yoshiyuki Matsuoka

Professor at Keio University, President of DesignJuku

Born in Shimonoseki, Yamaguchi, Japan. He is specialized in design science, design engineering, and product development system. In his specialized fields, he has proposed the bases of design science: multispace design model (M model), M method (multispace design method), AGE thinking model, emergent design, and many others. He has also proposed a new paradigm of design, which integrates timeaxis into designing, called "timeaxis design". Using the design theories and design methodologies he proposed, numbers of organizations have designed and created various products and systems.

He currently serves as the President of JSSD (Japanese Society for the Science of Design), the Vice-President of JSDE (Japan Society for Design Engineering), a Fellow of JFES (The Japan Federation of Engineering Societies), JSME (The Japan Society of Mechanical Engineers), a Director of SSDS (Society for the Science of Design Studies, Japan), and a board member of CG ARTS Society, Japan Industrial Designer Society, ASME, IEEE, Design Society. He also served as a visiting research fellow at Illinois Institute of Technology, a committee member of Ministry of Economy, Trade and Industry and The Small and Medium Enterprise Agency, a Director of traFST (Transdisciplinary Federation of Science and Technology), and a specialized judge of review committee for Machine Design Awards.

◆ 読者の皆さまへ ◆

平素より，小社の出版物をご愛読くださいまして，まことに有り難うございます．
　(株)近代科学社は 1959 年の創立以来，微力ながら出版の立場から科学・工学の発展に寄与すべく尽力してきております．それも，ひとえに皆さまの温かいご支援があってのものと存じ，ここに衷心より御礼申し上げます．
　なお，小社では，全出版物に対して HCD（人間中心設計）のコンセプトに基づき，そのユーザビリティを追求しております．本書を通じまして何かお気づきの事柄がございましたら，ぜひ以下の「お問合せ先」までご一報くださいますよう，お願いいたします．
　お問合せ先：reader@kindaikagaku.co.jp
　なお，本書の制作には，以下が各プロセスに関与いたしました：

- 企画：小山 透
- 編集：安原悦子，高山哲司
- 英文校正：山本優子，榮佑馬，有田実花子
- 組版デザイン：安原悦子
- 組版・印刷・製本：藤原印刷
- カバー・表紙デザイン：藤原印刷
- 広告宣伝・営業：山口幸治，冨髙琢磨，東條風太

日本語 - 英語バイリンガル・ブック

経営戦略に新価値をもたらす 10 の知恵
モノづくり × モノづかいの デザインサイエンス

© 2017　Matsuoka Yoshiyuki　　　　　　　　Printed in Japan
2017 年 12 月 31 日　初版第 1 刷発行

著　者　松岡由幸
発行者　小山　透
発行所　株式会社 近代科学社

〒162-0843　東京都新宿区市谷田町 2-7-15
電話　03-3260-6161　振替 00160-5-7625
http://www.kindaikagaku.co.jp

藤原印刷　　　ISBN978-4-7649-0525-2
定価はカバーに表示してあります．

・本文中における ©，®，™ 等の表示は省略しています．
・本書の複製権・翻訳権・譲渡権は株式会社近代科学社が保有します．
・JCOPY ＜(社) 出版者著作権管理機構　委託出版物＞
本書の無断複写は著作権法上での例外を除き禁じられています．
複写される場合は，そのつど事前に (社) 出版者著作権管理機構
(電話 03-3513-6969，FAX 03-3513-6979，e-mail: info@jcopy.or.jp) の許諾を得てください．

日本語－英語対訳本

Mメソッド －多空間のデザイン思考－

監修：デザイン塾
編著：松岡由幸
著者：氏家良樹・浅沼尚・髙野修治
　　　伊豆裕一・佐藤浩一郎・加藤健郎

A5判・カラー・160頁・2,200円+税

Mメソッド (Multispace Design Method) とは、デザイン要素を「抽出」、「分類」、「構造化」、「分解と追加」することで、従来では難しかった「自由な思考」と「理にかなった思考」の両立を実現する新しい思考メソッドである。本書では、このMメソッドを利用したデザイン手法を、具体的事例を用いながら、豊富な図とイメージにより解き明かす！
英語での思考方法も習得できるよう日英対訳で表記！